生命科学系列丛书

玉米苗期冷响应基因的克隆与功能鉴定

王晓宇 著

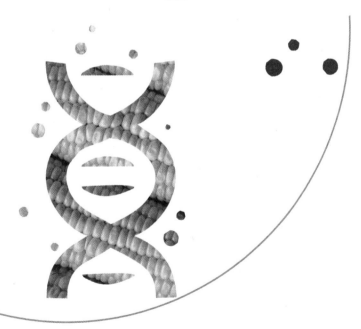

黑龙江大学出版社
哈尔滨

图书在版编目（CIP）数据

玉米苗期冷响应基因的克隆与功能鉴定 / 王晓宇著. -- 哈尔滨：黑龙江大学出版社，2019.6
ISBN 978-7-5686-0363-8

Ⅰ. ①玉… Ⅱ. ①王… Ⅲ. ①玉米－苗期－基因克隆－研究 Ⅳ. ①S513.051

中国版本图书馆 CIP 数据核字（2019）第 100767 号

玉米苗期冷响应基因的克隆与功能鉴定
YUMI MIAOQI LENGXIANGYING JIYIN DE KELONG YU GONGNENG JIANDING
王晓宇 著

责任编辑	张志雯
出版发行	黑龙江大学出版社
地　　址	哈尔滨市南岗区学府三道街 36 号
印　　刷	哈尔滨市石桥印务有限公司
开　　本	720 毫米×1000 毫米　1/16
印　　张	10.5
字　　数	161 千
版　　次	2019 年 6 月第 1 版
印　　次	2019 年 6 月第 1 次印刷
书　　号	ISBN 978-7-5686-0363-8
定　　价	32.00 元

本书如有印装错误请与本社联系更换。

版权所有　侵权必究

前　言

东北地区是我国玉米的主产区之一,其位于世界三大玉米黄金带上,由于昼夜温差大,非常适合玉米生长。然而由于地理条件等因素,玉米在苗期经常遭受低温侵害,产量下降。随着市场对玉米需求量的不断扩大,如何提高玉米产量成为当务之急。因此,选育抗冷玉米新种质和新品种,研究玉米的冷响应机制就显得尤为重要。

本书从转录组水平、抗冷相关基因功能分析等方面研究了玉米苗期冷响应的分子机制,取得了一些创新性成果。主要研究内容及成果如下:

1. 以玉米抗冷自交系 W9816 为试验材料,选取三叶期的叶片和根部组织,采用 cDNA 扩增片段长度多态性(cDNA amplified fragment length polymorphism,cDNA - AFLP)的方法,利用 174 对引物组合,比较低温处理前后基因的差异表达情况。根据 cDNA - AFLP 分析,在叶片中发现 6 829 个转录衍生片段(transcript - derived fragment,TDF),在根部发现 6 955 个 TDF。每对引物组合有 30 ~ 50 个 AFLP 扩增带,其大小为 70 ~ 600 bp。尽管大多数条带在冷胁迫后没有发生明显变化,但在叶片中仍然检测到 620 个差异表达基因(differentially expressed gene,DEG),在根部检测到 531 个 DEG。这些 DEG 显示出不同的表达模式。其中,仅 61 个上调的 DEG 和 32 个下调的 DEG 在两种组织中同时具有一致的表达趋势;大多数 DEG 冷胁迫后在两种组织中体现出不同的表达趋势。通过 NCBI BLASTX 和 MaizeGDB 数据库分析,根据功能将这些 DEG 分为 4 类——信号转导(10,15%)、转录调控(9,13%)、翻译和翻译后修饰(10,15%)、细胞代谢与组织(24,36%),还有 6 个 DEG 推测为编码蛋白(9%)、8 个 DEG 在数据库中没有匹配(12%)。同时利用实时荧光定量 PCR(real - time PCR)验证了 16 个与玉米抗冷相关的候选基因。

2. Sec14 类蛋白参与必要的生物学过程,例如磷脂代谢、信号转导、膜运输以及逆境响应。本书采用 cDNA 末端快速扩增(rapid amplification of cDNA end, RACE)PCR 技术成功克隆了一个磷脂酰肌醇转运相关蛋白基因 *ZmSEC14p*(登录号为 KT932998)。*ZmSEC14p* 基因包括一个完整的开放阅读框,推测可编码 295 个氨基酸。染色体定位表明 *ZmSEC14p* 基因定位于玉米基因组 1 号染色体上,横跨 3 420 bp,包括 5 个外显子。多重序列比对表明,玉米 ZmSEC14p 与高粱 SbSEC14p 氨基酸同源性最高;进化树分析表明,ZmSEC14p 属于未识别的同源组(uncharacterized SEC14p homolog group, UCSH)的一个成员,仅仅包含一个 SEC14 结构域。I-TASSER 结构预测表明,ZmSEC14p 蛋白包括 10 个 α 螺旋、5 个 β 折叠、3 个 3_{10} 螺旋,其中 7 个 α 螺旋、5 个 β 折叠以及 2 个 3_{10} 螺旋可形成磷脂结合口袋。基因表达模式研究发现,*ZmSEC14p* 基因转录受低温、盐以及脱落酸(abscisic acid, ABA)诱导表达;组织特异性分析表明,*ZmSEC14p* 在玉米叶片中表达最高;亚细胞定位表明,ZmSEC14p 蛋白主要定位于细胞核。对过表达 *ZmSEC14p* 转基因拟南芥植株在不同生长发育时期抗逆表型进行鉴定发现:在种子萌发阶段,转基因植株对低温的敏感性显著降低,对氯化钠和 ABA 的敏感性提高;在苗期,转基因植株在冷胁迫下具有更快的初生根生长速率;冷胁迫下,转基因植株较野生型植株的存活率提高了约 38%。活性氧组织定位以及抗氧化酶活性测定试验表明,冷胁迫后,相比于野生型株系,转基因株系过氧化物酶(peroxidase, POD)和超氧化物歧化酶(superoxide dismutase, SOD)的活性显著提高,酶活性的提高与 *ZmSEC14p* 调控抗氧化相关基因的表达有关。脯氨酸含量测定试验表明,低温处理 48 h 后,*ZmSEC14p* 转基因株系的脯氨酸含量分别是野生型株系的 1.35 倍和 1.61 倍。冷胁迫下,转基因拟南芥中一些逆境胁迫响应基因(如 *CBF3*、*COR6.6*、*RD29B*)的表达量较野生型显著上调。逆境响应基因的上调表达可能与 *ZmSEC14p* 调节磷脂酶 C(Phospholipase, PLC)基因的表达与酶的活性相关。

综合以上研究结果表明,*ZmSEC14p* 作为逆境胁迫响应基因在逆境调控网络中起正向调控作用,同时本书从转录水平进行研究,增加了对玉米冷响应机制的理解,为今后通过基因工程改良玉米抗逆性提供了重要的候选基因。

本书的出版得到了国家自然科学基金(31701467)、内蒙古民族大学博士科研启动基金(BS398)的资助,书中涉及的研究内容是笔者在吉林大学植物科学学院原亚萍教授的悉心指导下完成的,黑龙江大学出版社对本书的出版给予了大力支持,在此表示衷心的感谢。此外,内蒙古民族大学农学院李敏老师,内蒙古民族大学生命科学学院张继星教授,刘栩铭、卜祥琪、华文芝、由佳庆等同学为本书做了部分校稿工作,在此一并感谢。

由于本人水平有限,书中还有一些不尽如人意的地方,恳请读者批评指正。

笔　者
2019 年 4 月

目　录

第1章　文献综述	1
1.1　冷胁迫影响玉米产量	3
1.2　冷胁迫对玉米生长的影响	3
1.3　冷胁迫的整体效应	4
1.4　逆境响应蛋白	7
1.5　膜组分的改变	8
1.6　渗透调节物质	9
1.7　ROS清除系统	11
1.8　植物激素	12
1.9　低温信号转导途径	14
1.10　本研究的目的和意义	21
第2章　冷胁迫下玉米叶片和根部基因表达谱的研究	23
2.1　试验材料与方法	26
2.2　结果与分析	44
2.3　讨论	59
2.4　结论	61
第3章　玉米 $ZmSEC14p$ 基因的克隆及序列分析	63
3.1　试验材料与方法	66

  3.2 结果与分析……………………………………………… 74
  3.3 讨论……………………………………………………… 81
  3.4 结论……………………………………………………… 82
第 4 章 玉米 *ZmSEC14p* 基因的功能分析………………………… 83
  4.1 试验材料………………………………………………… 85
  4.2 试验方法………………………………………………… 89
  4.3 结果与分析……………………………………………… 103
  4.4 讨论……………………………………………………… 120
  4.5 结论……………………………………………………… 122
附录………………………………………………………………… 123
参考文献…………………………………………………………… 129

第1章

文献综述

玉米是世界上最重要的粮食作物之一。预计到2050年,全球将有大约97亿人口。到2050年,发展中国家对玉米的需求量将扩大一倍,因此对玉米需求的性质也随之发生改变。玉米不仅是一种重要的粮食作物,同时也是动物饲料的主要成分。在过去的几十年中,随着亚洲和拉丁美洲等地区经济的快速增长,人们生活水平不断提高,对家禽、家畜的需求增加,进而使得玉米饲料市场迅速扩大。玉米作为生产生物乙醇的关键成分,在工业原料方面也发挥着重要的作用。

玉米是我国三大粮食作物之一,东北地区素有我国的"玉米之乡"的美称,位于世界"三大玉米黄金带"上,由于昼夜温差大,非常适合玉米生长。玉米已经从初级食品(如淀粉、饲料)的初加工发展到更深层次产品(如玉米油、燃料乙醇、谷氨酸以及木糖醇等)的加工。玉米加工还涉及其他领域,例如纺织行业、汽车行业、医药行业以及材料行业等。由于玉米涉及的行业领域广,对玉米的需求量将不断扩大,为了提高单粒点播的产量,选育和创制高产、多抗的玉米杂交品种迫在眉睫。

1.1 冷胁迫影响玉米产量

非生物胁迫,例如冷胁迫、干旱和盐胁迫严重限制了作物的生长与产量。冷胁迫影响植物的生长、发育、空间分布和作物产量。玉米对冷比较敏感,冷适应性差。我国东北地区是玉米的主要产区,然而由于地理条件等因素,春季经常遭受冷害的侵袭,导致玉米大面积减产。

1.2 冷胁迫对玉米生长的影响

玉米起源于亚热带,对低温非常敏感。低温可以影响玉米的萌发阶段、营养生长阶段以及生殖生长阶段。

低温条件下,种子萌发的时间与低温强度成正相关。温度越低,种子的萌发率越低。种子吸水后,若长期处于低温条件下,可增加土壤病原菌入侵的机会。不同基因型玉米在种子萌发阶段对低温的耐受能力不同。6 ℃冷胁迫下,相比于感冷和中度耐冷基因型,耐冷基因型玉米具有更高的萌发率,原因可能在于冷胁迫下耐冷玉米受到的吸胀损伤小、α-淀粉酶活力和脱氢酶活力较高。

在玉米营养生长阶段,冷胁迫可导致植株明显矮小,主要表现为叶、茎等生长发育缓慢、光合作用能力降低,进而减少干物质的积累。低温可减少玉米的叶数和延缓叶片的伸出速度。W. G. Duncan 等人发现当昼夜温度从 36 ℃/31 ℃降到 15 ℃/10 ℃时,玉米平均叶数从 26 片降到 19 片;当茎生长点处于 12~26 ℃范围内时,叶片伸出速度与温度呈线性关系。低温降低了光合速率,导致玉米幼苗干物质积累下降。有研究表明,低温持续 10 d,全株干物质积累可降低 21.4%。低温可以影响根系的生长状况和活跃程度。例如,冷胁迫下,玉米根伸长区肿大呈鸡爪状,细胞分裂受到影响。低温可以降低玉米的根系活力,进而抑制根系的呼吸作用,减少供给植物生长所需的能量。低温可影响根系形态以及降低根系导水率,强烈抑制根系对矿物离子和水分的吸收,植株表现出萎蔫症状。

玉米生殖生长阶段遭遇低温必然会造成减产。孕穗期的玉米遭受低温可抑制雌穗分化,导致败育花的增加,最终使产量下降。玉米在抽雄到开花期遇到 18 ℃低温可延长雌穗吐丝期、缩短雄穗开花持续时间,造成花期不遇。玉米在灌浆期遭遇低温(20 ℃)时,籽粒灌浆速度减慢;温度降至 18 ℃时,灌浆速度明显减慢,16 ℃时灌浆过程基本停止。低温严重影响灌浆期玉米籽粒的灌浆过程,进而导致灌浆较晚,霜前不能正常成熟,籽粒含水量高、成熟度下降,最终导致减产。

1.3 冷胁迫的整体效应

低温会直接影响许多球状蛋白质的稳定性和可溶性,进而干扰蛋白质复合物的稳定性以及代谢调控。低温可降低膜的流动性,膜易发生固化现象,导致所有有膜参与的生物学反应发生紊乱,例如离子通道的开闭、电子传递反应等。低温可增加 ROS(活性氧类)的积累,降低清除活性氧相关酶的活性。ROS 的积累对细胞,尤其是膜,将产生有害的影响,如离子渗漏。另外,低温可增加 RNA 二级结构的形成,影响基因和蛋白质的表达。

光合作用可将光能转化为活跃的化学能,储存于高能中间产物 NADPH

和 ATP 中,同时利用 NADPH 和 ATP 将 CO_2 还原成糖类,将活泼的化学能转化为稳定的化学能,储存于光合产物中。卡尔文循环是主要储存化学能量的代谢过程。CO_2 被五碳化合物 RuBP 固定产生甘油酸-3-磷酸,进而借助 ATP 和 NADPH 的能量,在 3-磷酸甘油酸激酶和甘油醛-3-磷酸脱氢酶的作用下,产生甘油醛-3-磷酸。大多数甘油醛-3-磷酸留在叶绿体中再生 RuBP,一部分被运输到细胞质中用于细胞内代谢或是转化成蔗糖。蔗糖合成过程中释放无机磷,然后被运往叶绿体合成 ATP(图 1-1)。

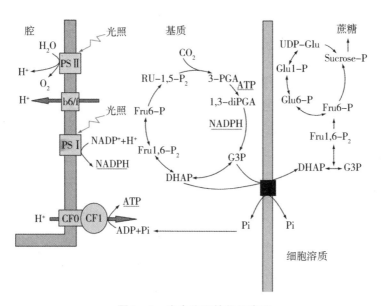

图 1-1 光合作用简化示意图

注:b6/f 为细胞色素 b6f 复合体;CF0 和 CF1 为偶联因子;ADP 为二磷酸腺苷;Pi 为磷;DHAP 为二羟丙酮磷酸;Fru1,6-P_2 为果糖-1,6-二磷酸;Fru6-P 为果糖-6-磷酸;RU-1,5-P_2 为核酮糖-1,5-二磷酸;3-PGA 为甘油酸-3-磷酸;1,3-diPGA 为甘油酸-1,3-二磷酸;G3P 为三磷酸甘油醛;UDP-Glu 为 UDP-葡萄糖;Glu1-P 为葡萄糖-1-磷酸;Glu6-P 为葡萄糖-6-磷酸;Sucrose-P 为蔗糖磷酸。

低温可影响光合作用的各个方面。例如,低温抑制细胞内蔗糖的合成,导致磷酸化中间产物的积累,从而消耗可用无机磷以及降低无机磷在细胞质与叶绿体之间的循环。另外,低温增加了膜的黏度,影响了质体醌的扩散,进而抑制了类囊体电子传递链。相反,由PSⅠ和PSⅡ天线分子捕获光能以及将光能传递到反应中心驱动电荷分离反应等过程大多数对温度不敏感。由于低温导致光合作用反应的不平衡,类囊体膜处于超能状态,结果增加了ROS的积累。

玉米叶片植入人工电子受体的试验表明,温度对光合作用的光反应影响很小。低温通过降低酶反应速率来影响玉米光合作用的暗反应。例如,A. H. Kingston – Smith 等报道玉米苹果酸脱氢酶和核酮糖-1,5-二磷酸羧化酶冷胁迫下活力很低。

正常生长条件下,ROS清除系统(如 Cu – Zn SOD 和 APX)可以维持由光合电子传递链产生的活性氧水平。低温条件下,这些酶系统的活性较低,同时PSⅠ还原力的积累会导致光抑制和膜脂过氧化。O_2^- 自由基是引起PSⅠ光损伤的主要因素,这类ROS可攻击铁硫中心。低温、低光强下,PSⅠ光抑制不仅发生在冷敏感植物(如黄瓜、番茄等)中,同时也发生在耐冷植物(如大麦、冬黑麦以及拟南芥)中。S. E. Tjus 等指出造成植物抗冷感冷的原因之一是清除活性氧相关酶对低温敏感程度不同以及耗散多余激发光能的方式存在差异。PSⅠ产生的活性氧导致PSⅠ的光抑制,如果不能及时清除这些活性氧,那么它们可通过 O_2^- 歧化酶或者自动歧化作用产生 H_2O_2。这些 H_2O_2 和 O_2^- 能够扩散和抑制PSⅡ。

除了光抑制对两个系统的影响之外,ROS的形成还会导致类囊体膜脂过氧化。由于脂中包含高比例不饱和脂肪酸的半乳糖(18:3),叶绿体膜易受活性氧的光破坏。ROS的形成会消耗一部分叶绿体氧化还原物质,打破了氧化还原平衡。ROS对硫氧还蛋白的还原性以及其通过还原二硫键激活靶酶的能力具有负调控效应。对水稻、拟南芥低温条件下蛋白质组学的研究表明,低温诱导了光合相关蛋白,例如 RcbA、PSⅡ放氧复合体蛋白、景天庚酮糖二磷酸酶、ATP 合酶 a 链和 b 链。这些蛋白质及蛋白质复合体的降解可能与低温下产生的 ROS 有关。在感冷黄瓜完整的叶片中,羟基自由基可对核酮糖-1,5-二磷酸脱羧酶大亚基进行位点的特异性切割。

1.4 逆境响应蛋白

1.4.1 COR 和脱水蛋白

冷胁迫可以增加亲水蛋白的积累。许多编码这些蛋白质的基因参与低温、干旱和 ABA 响应。大多数基因命名为 COR、LTI、RAB、KIN、ERD。小麦脱水蛋白 WCOR410 的积累能提高植物的抗冻性。低温驯化期间,耐冻冬小麦品种"Fredrick"相比中度耐冻春小麦品种"Glenlea"能够积累更多的 WCOR410。但是,单一过表达脱水蛋白不一定会增加植株的抗冻性。例如,过表达冷诱导脱水蛋白基因 *RAB*18 对提高拟南芥植株的抗冻性没有影响。尽管如此,仍有多个脱水蛋白基因的过表达能提高植株的抗冻性。*RAB*18 和 *COR*47 或者 *LTI*29 和 *LTI*30 在拟南芥中同时过表达,抗冻试验表明相比于野生型植株,转基因植株降低了离子渗漏。目前,脱水蛋白提高植株抗冻性的机制还不太清楚。小麦 WCOR410 的积累发生在细胞膜的附近,而拟南芥 LTI29 在细胞质、细胞膜以及液泡膜均有积累。低温处理后,LTI29 优先定位于细胞膜;相反,LTI30 特异性定位于膜。定位试验表明,在脱水逆境中,脱水蛋白可以保护膜的稳定性。事实上,脱水蛋白中两亲螺旋的存在可以作为膜和细胞质之间的接口。玉米脱水蛋白 DHN1 可以结合含有酸性磷脂的脂质囊泡。

脱水蛋白不仅能够维持细胞膜的稳定性,还具有冷冻保护活性。温州蜜柑脱水蛋白 CuCOR19 可以作为羟基自由基清除剂。体外试验表明,相比于血清血蛋白、还原型谷胱甘肽、脯氨酸或者甜菜碱,CuCOR19 能够更高效地防止大豆脂质体过氧化。过表达 *CuCOR*19 提高了转基因烟草植株的抗冻性,其抗冻性与 CuCOR19 积累的水平呈正相关。

冷胁迫下,脱水蛋白仅是亲水蛋白的一个亚族。COR15 属于亲水蛋白,定位于叶绿体气孔,表达受低温高度诱导。体外异源表达 *COR*15 能提高非驯化植株叶绿体的抗冻性。抗冻性的提高可能是由于 COR15 对基质蛋白起到冷冻保护的作用。

1.4.2 AFP

为了防止细胞核结冰和晶体融合,越冬植物能够分泌 AFP 蛋白。AFP 被分泌到质外体,不可逆结合到冰的表面。AFP 的活性受低温诱导。冷胁迫下,冬黑麦 AFP 诱导积累于叶片和冠部质外体。体外试验表明,从冷适应冬黑麦叶片提取的质外体蛋白能够抑制冰的再结晶,同时也能够减慢冰移动的速度。植物中 AFP 可能是通过与冰直接作用以及减慢冰的再结晶发挥作用的。

有研究人员预测,多年生黑麦草的 AFP 在蛋白相反位置存在两个冰结合结构域。大约 74% 的胡萝卜 AFP 由一个富含亮氨酸的重复序列组成。小麦的两个 AFP 分别包括 2 个或 8 个富含亮氨酸的重复序列。冬黑麦脂转运蛋白 1 具有抗冻活性。但是,大多数植物 AFP 属于 PR 蛋白,包括几丁质酶、β-1,3-葡聚糖酶、索马甜等相关蛋白。冬黑麦 AFP 具有抗真菌活性。但在常温下,水杨酸刺激积累的 PR 蛋白不具有抗冻活性。

1.4.3 CSP 和 RNA 结合蛋白

冷胁迫下,植物面临的主要问题之一是 RNA 二级结构的形成与稳定。事实上,细菌 CSP 具有 RNA 分子伴侣的活性。这种功能对低温下 mRNA 的有效翻译非常关键。大肠杆菌 CSPA 是 9 个大肠杆菌蛋白中最突出的蛋白,低温条件下,其含量可占总蛋白质的 10%。其他 RNA 结合蛋白在低温驯化过程中具有重要的作用。冷胁迫后,拟南芥 *AtGPR2* 的转录水平显著上升。低温(11 ℃)处理后,过表达 *GPR2* 转基因拟南芥植株的根部生长速度较野生型快;-5 ℃ 处理 2 h 后,大多数野生型和突变体植株死亡,70% ~ 80% 的 *GPR2* 转基因植株存活。拟南芥 *atRZ-1a* 的转录水平受低温显著诱导,低温处理下,*atRZ-1a* 突变体植株的种子萌发和幼苗生长相比野生型植株受到显著抑制;相反,过表达 *atRZ-1a* 可促进植株种子萌发和幼苗生长。

1.5 膜组分的改变

低温诱导后膜组分发生重要变化。例如,在冬黑麦和拟南芥的叶绿体中,低温驯化可导致内外膜单酰基甘油的含量降低,二半乳糖基二酰甘油的含量增加。低温驯化的拟南芥细胞膜磷脂增加,脑苷脂和游离甾醇含量降低。低温驯化下,番茄品种抗冻性的提高与细胞质膜磷脂的增加相关,主要是磷脂酰乙醇胺含量增加。拟南芥低温驯化期间,不同磷脂间的比例差异很小,但是磷脂内组分之间的差异很大。脂的不饱和程度与低温响应有关。拟南芥 $fad2$ 突变体在微粒体中不具有 18:1 去饱和酶的活性。突变体植株降低了多不饱和脂肪酸的水平:$fad2.2$ 突变体植株中磷脂酰胆碱含有 2% 的 18:2 脂肪酸和 14% 的 18:3 脂肪酸,而野生型植株中这两种脂肪酸的含量分别为 33% 和 40%。突变体与野生型在 22 ℃ 下生长状况相似,但是当转移到 6 ℃ 时,突变体的莲座叶逐渐死亡。冷胁迫下,耐冷马铃薯细胞膜中 16:0 到 16:2 的极性脂的含量增加,感冷马铃薯中观察不到此现象。耐冷油菜品种相比感冷油菜能迅速积累更多的 18:3 不饱和脂肪酸。

不饱和脂肪酸增加的原因之一可能是用于缓解低温诱导的膜固化,因为不饱和脂肪酸能使膜处于液态。甘蓝型油菜膜的流动性与抗冷性的增强相关。但是,G. Tasseva 等报道膜的流动性增加不足以弥补由低温导致的膜的固化作用。冷胁迫下,多不饱和脂肪酸对维持叶绿体的稳定性起重要作用。拟南芥叶绿体 $\omega-3$-去饱和酶基因 $FAD8$ 的转录水平受低温强烈诱导。低温是叶绿体膜从液晶相到凝胶相转变的影响因素。三烯酸的诱导可以防止叶绿体膜凝胶相的形成。

1.6 渗透调节物质

除了可溶性糖,相容性分子还指包含氨基酸分子(丙氨酸、甘氨酸、丝氨酸)、多胺和甜菜碱的异质组。相容性分子是低分子质量的有机分子,响应多种逆境,如渗透和冷胁迫。生理学上,即使高浓度的相容性物质也不会产生不良代谢作用。但是,它们响应逆境的方式还不是非常明确。

1.6.1 脯氨酸

低温可促进脯氨酸积累,脯氨酸能提高植株的抗冷性。例如,4 ℃ 低温处理 4 h 后,拟南芥脯氨酸的含量是常温下的两倍。ProDH 催化降解脯氨酸,

*AtProDH*反义转基因植株相比野生型能够积累更多的脯氨酸。相比于野生型植株，*AtProDH*反义转基因植株抗冷性更强，这表明脯氨酸含量和抗冻性之间存在正相关。但是，不同拟南芥品种之间的杂交试验表明抗冻杂种优势效应与脯氨酸含量之间不存在显著性关系，脯氨酸含量的差异对拟南芥抗冻杂种优势的建立没有发挥重要作用。因此，即使冷胁迫下积累脯氨酸，其含量也不能作为解释拟南芥不同品种之间抗冻能力的关键因素。脯氨酸促进植物抗冷的机制还不清楚。脯氨酸作为相容性溶质，在防止酶变性、稳定蛋白质、调控细胞酸度以及增加水的结合能力中扮演着重要的角色，同时可作为碳源和氮源的储存器。

1.6.2 甜菜碱

甜菜碱是季胺化合物，在叶绿体中由胆碱的两步氧化合成。胆碱单加氧酶催化胆碱合成甜菜碱醛，甜菜碱醛通过醛脱氢酶合成甜菜碱。并不是所有的植物在响应非生物胁迫时均积累甜菜碱。例如，菠菜和小麦自然积累甜菜碱，而拟南芥、番茄、马铃薯和水稻不积累甜菜碱。冷胁迫下，甜菜碱可能稳定转录和翻译蛋白复合物和膜。此外，甜菜碱可以间接诱导 H_2O_2 介导的信号通路。植物中 H_2O_2 可以作为第二信号分子。高浓度的 H_2O_2 可导致细胞程序性死亡，低浓度的 H_2O_2 可以调控基因表达，增加植物抗逆性。这是因为低浓度的 H_2O_2 能够诱导抗氧化酶如 CAT 的表达从而提高植物抗冷性。外施甜菜碱或者 H_2O_2 能够提高野生型番茄的抗冷性。甜菜碱能够帮助形成 H_2O_2，推测外施甜菜碱提高植物的抗冷性可能是诱导了 H_2O_2 介导的抗氧化机制。

1.6.3 多胺

多胺是具有两个或更多伯氨基的有机化合物。多胺生物合成途径起始于腐胺，然后在两个氨丙基转移酶（亚精胺合成酶和精胺合成酶）连续作用下分别转化为亚精胺和精胺。这两种酶都利用 SAM 作为氨丙基的供体。冷胁迫可上调水稻根部亚精胺合成酶基因 *OsSPDS2* 的表达。拟南芥 4 ℃处理 14 d，植株体内腐胺、鸟氨酸、瓜氨酸以及多胺前体的含量均上升。水稻处于 5 ℃时，耐冷品种"Yukihikari" *OsSAMDC* 的转录水平持续上升到 72 h，但在感冷籼稻品种"TKM9"中没有积累。有试验表明多胺提高抗冷的作用与光合功能相关。多胺与类囊体膜有关，尤其是 LHC II 和 PS II。四季豆冷胁迫(6 ℃)52 h 可导致 LHC II 相关腐胺的降低以及精胺的增加。另外，多胺在缓解氧化逆境中扮演重要的角色。植物处于低温逆境下，抑制多胺的合成将增加氧化破坏，例如电渗漏、降低 PS II 的光合效率。过表达 *SAMD* 转基因烟草能够积累高水平的可溶性多胺，进而缓解低温诱导的损伤。逆境处理的转基因烟草相对于对照组，抗氧化酶基因的转录水平显著被诱导，例如 *APX*、*SOD* 和 *GST*。多胺积累是否能够提高抗氧化酶基因的转录水平还有待考证。多胺可能通过提高转录因子 DNA 结合活性作为信号分子。

1.7 ROS 清除系统

ROS 在叶绿体、线粒体、过氧化物酶体和细胞质中形成。线粒体中，ROS 能够通过 UQ/UQH2 循环产生。低温环境可以积累 ROS。为了平衡 ROS 的水平，植物诱导和激活 ROS 清除系统，例如 CAT 和 SOD 活性的上升。拟南芥 4 ℃ 处理 14 d 后，抗坏血酸的含量增加。抗坏血酸是植物丰富的抗氧化剂。叶绿体中 H_2O_2 的解毒只依赖于 PS I 附近结合到类囊体膜上的 APX 的活性。单脱氢抗坏血酸还原酶在消耗 NADPH 的情况下，催化叶绿体中抗坏血酸的再生。GSH 在大多数组织、细胞和高等植物的亚细胞区室均有发现。GSH 主要以还原形式存在，能够与单线态氧、超氧化物和羟基自由基相互作用，因此是一种自由基清除剂。GSH 可以通过去除由脂质过氧化产生的酰基过氧化物来稳定膜结构。

在植物中，抗坏血酸氧化酶对维持合适的 ROS 浓度起着重要的作用，同时基因的表达受低温诱导。冷胁迫下，铁蛋白和 GST 的蛋白丰度上调。铁蛋白保护植物免遭非生物胁迫引起的氧化损伤。GST 催化 GSH 与一系列氧化逆境产生的有毒底物结合，进而起到解毒作用。拟南芥低温驯化过程中可溶性蛋白质组的研究识别了许多受低温调控且参与氧化还原调节和 ROS 清除的蛋白质。

抗冷与 ROS 清除系统的激活存在一定的相关性。8 ℃ 处理 5 d，水稻耐冷品种"Xiangnuo No.1"和"Zimanuo"相比感冷品种"Xiangzhongxian No.2"和

"IR50"具有更少的电渗漏和更低的 H_2O_2 含量。低温处理 3 d 后,抗氧化酶(SOD、CAT、APX)的活性和抗氧化物(抗坏血酸和还原型谷胱甘肽)的含量在两个耐冷品种中显著增加,而在感冷品种中显著降低。这些结果表明水稻耐冷与低温条件下抗氧化系统能力的提高有关。

蛋白质甲硫氨酸残基是 ROS 氧化的主要靶标。甲硫氨酸亚砜还原酶是抗氧化酶,参与甲硫氨酸亚砜到甲硫氨酸的转变。拟南芥甲硫氨酸亚砜还原酶的基因 *MsrB3* 是冷响应基因,*MsrB3* 突变体植株在低温驯化后失去了抗冻的能力。

1.8 植物激素

1.8.1 ABA

ABA 在低温响应中扮演重要的角色。拟南芥 *ABA3* 突变体植株相比野生型植株,抗冷性降低的同时 ABA 含量下降。ABA 可以提高植物的抗冷性,低温可诱导 ABA 的产生。低温(白天 4 ℃/夜晚 2 ℃)处理 15 h 后,拟南芥 ABA 含量增加了 4 倍,处理 24 h 后,其含量恢复到对照水平。干旱处理 3 h,植株 ABA 含量增加 20 倍。因此,低温促进 ABA 含量的上升较为适中。这些研究表明 ABA 在响应干旱胁迫的过程中更为重要。低温(0 ℃)处理 3 h、6 h 和 24 h 后发现拟南芥中已知 ABA 合成相关基因并不受低温调控。因为 ABA 的合成主要受转录水平调控,试验数据表明 ABA 合成在早期低温响应中并不是一个重大事件。此外,一些冷响应基因(*LTI78*、*ADH*、*UGPase*)在 ABA 缺陷型突变体中仍然被诱导表达。但是,低温诱导下,一些基因(*RAB18*、*RCI2A*)的表达是依赖于 ABA 的。因此,低温驯化似乎主要不依赖于 ABA,但是 ABA 可能促进植物最大抗冷性或抗冻性的获得。事实上,许多冷响应基因的启动子区存在 ABRE 或是 DRE/CRT 顺式作用元件,这些基因能够被 ABA 诱导的蛋白所激活。一个 bZIP 类转录因子(ABF1)能够结合 ABA 响应元件,早期冷胁迫可持续诱导 *ABF1* 转录。*ABF4/DREB2* 也受低温诱导。因此,冷胁迫下,ABF1 和 ABF4 可能调控一些 ABA 诱导基因的表达。

1.8.2 水杨酸

低温能够提高水杨酸的含量。4 ℃处理拟南芥代谢组的研究表明,水杨酸在冷胁迫 1 h 开始积累,4 h 和 12 h 达到高峰,24 h 降低,然后持续增长到 96 h。5 ℃处理拟南芥芽能够积累水杨酸。过表达 *NahG* 基因能够在 5 ℃下积累水杨酸,随后由于羟化酶活性的降低,水杨酸含量下降。过表达 *NahG* 转基因植株在常温下生长速率与野生型植株相似,低温下生长速率减慢。但是,转基因 *NahG* 植株相比野生型植株具有更快的相对生长速率。

1.8.3 油菜素内酯

油菜素内酯是一类甾体化合物,结构类似植物和昆虫甾体激素。油菜素内酯控制广泛的生物学响应,例如细胞分裂、细胞扩张、营养生长和顶端优势。*CPD/DWF3* 基因编码细胞色素 P450 90A 家族的一个成员,在油菜素内酯合成过程中负责长春花甾酮 C-23 位继续羟基化形成茶甾酮,其在低温处理 24 h 后下调表达。推测绿豆 *CYP90A2* 基因参与油菜素内酯合成,其序列与拟南芥 *CPD* 基因的同源性高达 77%。*CYP90A2* 基因表达显著受冷抑制。低温导致生长停滞可能与油菜素内酯含量的下降有关。外施油菜素内酯可使由低温导致生长抑制的植株得以恢复。事实上,低温处理抑制绿豆上胚轴的生长后用 24-表油菜素内酯处理,可部分恢复上胚轴的生长。另外,低温(2 ℃)处理下,24-表油菜素内酯处理拟南芥植株相比不处理而言,*CBF*1、*LTI*178 和 *COR*47 基因的表达量上升得更快。

1.8.4 茉莉酸

丙二烯氧化物环化酶是茉莉酸生物合成途径的第一个关键酶。低温可上调此过程的基因的表达。拟南芥低温处理 10 d 后,叶间质丙二烯氧化物环化酶基因的表达明显增加。脂氧化酶参与茉莉酸生物合成途径,盐芥基因芯片数据表明低温诱导脂氧化酶基因的转录。因此,低温可诱导茉莉酸含量的上升。

1.9 低温信号转导途径

冷胁迫能够提高基因转录、蛋白质合成以及不同 ROS 清除酶的活性,也可增加细胞单线态氧、H_2O_2 和 O_2^- 自由基的含量。冷害的表型症状包括种子萌发率低、苗发育不良、叶片萎黄、叶片扩张降低,最终导致组织死亡。低温条件可能引起细胞脱水,导致膨压降低。膨压引起渗透势的改变可能是分子水平激活冷响应的主要原因。低温首先被细胞膜上的受体感知,信号被进一步传递并引起冷响应基因的表达。了解抗冷机制以及参与抗冷代谢网络的基因对作物改良意义重大。许多基因既参与冷胁迫,又参与干旱胁迫,表明在低温和干旱条件下植株存在相互交叉的代谢途径。

1.9.1 逆境感受器

干旱、盐和冷胁迫能够诱导钙离子瞬时通过质外体空间和内存库流入细胞质内。钙离子通道代表一类信号感受器,配体敏感的钙离子通道控制细胞内钙离子的释放。膜定位类受体蛋白激酶可控制各种信号事件。多种植物(如拟南芥、水稻、苜蓿和大豆等)内已发现含有参与感受环境信号和非生物胁迫信号的类受体蛋白激酶。类受体蛋白激酶接收外界信号并经由细胞内丝氨酸/苏氨酸激酶结构域激活下游代谢途径。结构上,它们由胞外区、一个跨膜区和细胞内激酶区组成。组氨酸激酶是一类膜定位激酶,参与感知渗透胁迫和植物非生物胁迫。冷胁迫下,微管可作为信号分子。微管解聚现象通常能在冷胁迫下观察到。微管与冷胁迫下 ABA 信号通路密切相关。多个感受器可能同时感受低温信号,每个感受器都可以感知一个逆境的特定方面以及参与低温信号传导的不同分支。

1.9.2 信号转导

细胞膜具有流动性,低温能降低其流动性,增加膜的固化。冷胁迫可以诱导植物细胞膜的流动性、蛋白质和核酸的构象改变,或是代谢物浓度的变化。紫花苜蓿和油菜药理学试验表明传递质膜固化能够诱导冷响应基因的表达。

二级信号(如 ABA 和 ROS)能够影响钙离子传递,进而影响低温信号。拟南芥 *aba*3/*frs*1(*LOS*5)突变体植株对冻害非常敏感,*LOS*5 突变体植株可以显著降低冷和渗透胁迫诱导的基因表达。各种非生物胁迫导致 ROS 在细胞内积累,ROS 对低温调控基因的表达具有深远的影响。拟南芥 *fro*1 突变体植株的 ROS 在细胞内高度积累,抑制冷响应基因的表达,植株表现为对冷害和冻害非常敏感。*FROS*1 编码线粒体呼吸电子传递链复合体Ⅰ的 Fe‑S 亚基,基因突变导致高水平 ROS 的产生。ROS 除了影响钙离子信号,还可直接通过激活氧化还原响应蛋白(如转录因子和蛋白激酶等)发挥作用。

冷胁迫无论是直接通过降低生化反应的速率还是间接通过基因表达重编辑都极大影响了细胞的新陈代谢。除了作为渗透保护剂和渗透剂,低温驯化诱导的特定代谢产物还可能作为信号,重新配置基因的表达。例如,芯片和 RNA 印记分析表明脯氨酸能够诱导许多启动区含有脯氨酸响应元件基因的表达。

1.9.3 转录调控

1.9.3.1 ICE1‑CBF 转录级联反应

CBF 通过结合冷响应基因启动子顺式作用元件激活基因的表达。转基因植株分析表明,异源表达 *CBF* 基因能够激活 *COR* 基因的表达。CBF 参与肌醇磷脂代谢、转录、渗透调节物质合成、ROS 解毒、膜转运、发育、激素代谢和信号转导等的调控。抗冷植物(如小麦、大麦和油菜)和感冷植物(如水稻、番茄和樱桃)中已经克隆到了 *CBF* 同源基因。在不同植物中过表达拟南芥 *CBF* 基因能够提高植株抗冷性或抗冻性;相反,其他植物 *CBF* 基因在拟南芥中异源表达能够提高转基因植株的抗冻性。因此,CBF 在不同植物低温驯化过程中扮演着重要的角色。冬季,植物在组成型抗冻和后天抗冻方面表现出显著的基因型差异。这两种特征似乎存在单独的遗传学调控。但是,组成型抗冻的分子机制还不明确。对组成型抗冻之间存在差异的拟南芥种质的转录组和代谢组研究表明,CBF 途径可能在组成型抗冻中扮演重要角色。在拟南芥中,ICE1 是 MYC 型碱性螺旋‑环‑螺旋转录因子,可以结合 *CBF*3 启动子区 MYC 识别元件,在低温驯化过程中对 *CBF*3 的表达起着重要的调控作用。*ICE*1 突变体植株不能诱导 *CBF*3 的表达,结果导致植株对冷害非常敏感。低温驯化过程中,组成型过表达 *ICE*1 能够提高 *CBF*3、*CBF*2 和 *COR* 基因的表达,同时能够提高转基因植株的抗冻性。*ICE*1 组成型表达且定位于细胞核,但只有在低温条件下诱导 *CBF* 的表达。这表明低温诱导 *ICE*1 翻译后修饰对激活下游基因表达是必要的(图 1-2)。事实上,低温可诱导 ICE1 磷酸化。转录组分析表明显性 *ICE*1 突变体植株阻碍 40% 的冷调控基因表达。参与钙离子信号、脂信号或者编码类受体蛋白的冷诱导基因受 *ICE*1 突变体的影响。拟南芥 *ICE*1 突变体植株对冷非常敏感。在非逆境条件下,*ICE*1 突变体影响 204 个冷响应基因的基本转录水平。这些基因基本的表达对拟南芥抗冷非常重要,因为 *ICE*1 突变体中这些基因表达的改变与感冷相关。

1.9.3.2 *CBF* 的负调控者

调控冷响应基因表达的转录因子的反馈抑制在维持最佳冷诱导转录组中起着重要的作用。拟南芥 *cbf*2 无义突变分析表明,低温驯化过程中,CBF2 是 *CBF*1 和 *CBF*3 表达的负调控者。相反,CBF3 可能调控 *CBF*2 的表达,因为 *ICE*1 突变体中 *CBF*3 表达的降低伴随着 *CBF*2 表达的上升。拟南芥 *CBF* 受上游转录因子 MYB15 的负调控。MYB15 能够结合 *CBF* 启动子区 MYB 识别元件。低温驯化过程中,*MYB*15 T-DNA 敲除突变体植株能够提高 *CBF* 的表达,从而提高植物的抗冻性。转基因植株过表达 *MYB*15 能够降低 *CBF* 的表达,植物的抗冻性随之下降。因此,MYB15 是 *CBF* 表达的上游负调控转录因子。有趣的是,相比野生型而言,冷胁迫下,*ICE*1 突变体中 *MYB*15 的转录本增多,因此 ICE1 能够负调控 *MYB*15 的转录水平。酵母双杂交及拉下试验表明 MYB15 能够与 ICE1 相互作用,但是 ICE1-MYB15 在低温驯化过程中相互作用的功能性意义还不清楚。

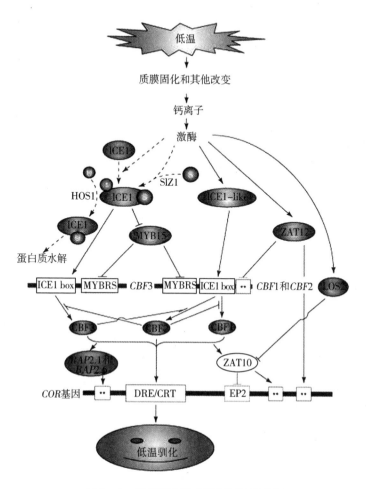

图 1-2 拟南芥冷响应转录调控示意图

在拟南芥中,冷诱导 C_2H_2 锌指转录因子基因 ZAT12,为 CBF 的负调控者。过表达 ZAT12 转基因植株在低温下能够降低 CBF 的表达。对拟南芥 LOS2 突变体的分析识别了另一个 C_2H_2 锌指蛋白基因——ZAT10/STZ,其可能作为 CBF 靶基因的负调节者。LOS2 是一个双功能烯醇化酶,能够在体外结合 ZAT10 启动子区 MYC 识别元件,LOS2 突变体植株在低温处理后能够诱导 ZAT10 的转录。瞬时表达试验表明 ZAT10 能够抑制 RD29A 的表达。CBF 在介导或调节冷诱导 ZAT10 中扮演重要的角色。ZAT10 和 ZAT12 可能在非生物胁迫转录调控网络中充当节点的角色,因为这些基因在受冷诱导的同时,也受其他非生物胁迫诱导,过表达这些基因的转基因植株渗透和氧化胁迫耐性有所提高。

1.9.3.3　不依赖于 CBF 的转录调控

芯片分析表明 CBF 仅调控 12% 的冷响应基因。因此,非 CBF 转录因子可能调控大部分冷响应基因。在大豆中,冷诱导 C_2H_2 锌指蛋白 SCOF1 能够诱导 *COR* 基因的表达。拟南芥 *ESK*1 突变体组成型积累高水平的脯氨酸,植株表现组成型抗冻。*ESK*1 编码一个 DUF231 蛋白。低温不能改变 *ESK*1 的表达。转录组分析表明在 *ESK*1 突变体中,312 基因的表达受影响,其中仅有 12 个基因在 *ESK*1 突变体植株和 *CBF*2 过表达植株中同时上调表达。因此,隐性 *ESK*1 突变导致植株抗冻的机制可能不同于 CBF 调控低温驯化的机制。通过采用下调 P_{RD29A}::*LUC* 报告基因表达的遗传学筛选方法,两个组成型表达的转录因子 HOS9 和 HOS10 被识别。尽管冷诱导 *CBF* 的表达与野生型相似,但 *HOS*9 突变体在低温驯化前后均体现出对冷敏感。芯片分析表明 HOS9 具有不同于 CBF 的调控基因。尽管低温下一些冷响应基因上调表达,但 *HOS*10 - 1 突变体植株不抗冻。HOS10 正调控 *NCED*3 基因的表达,因此 HOS10 可能调控 ABA 介导的低温驯化。

1.9.4　转录后调控

除了转录水平调控,基因表达受前体 mRNA 加工、mRNA 稳定性、在核内的运输以及翻译等转录后调控。最新研究表明转录后调控在低温驯化过程中扮演重要的角色。

1.9.4.1　RNA 加工和细胞核内的运输

前体 mRNA 剪切对功能性 mRNA 的合成具有重要的意义。响应发育与环境信号的可变剪切能够使细胞合成来自同一个基因的不同蛋白质。小麦两个早期 *COR* 基因(*ribokinase*、$C_3H_2C_3$ *RING - finger protein*)受冷调控,其成熟 mRNA 里保留了内含子。通过使用 P_{RD29A}::*LUC* 遗传学筛选,B. H. Lee 等识别了 *STABILIZED*1(*STA*1)基因——一个核前体 mRNA 剪接因子。*STA*1 作为前体 mRNA 剪切的调控者,对拟南芥抗冷具有重要的作用。低温能够上调

STA1 的表达。STA1 突变体植株不能对冷诱导基因 COR15a 前体进行剪切,导致植株对冷、ABA 和盐胁迫非常敏感。SR 蛋白是剪切体的一部分,能够作为真核生物的剪切调控者。在拟南芥中,冷和热胁迫能够调节许多 SR 蛋白基因前体 mRNA 的可变剪切,结果可能产生具有不同可变剪切功能的 SR 蛋白亚型。环境信号传导到核内改变基因转录,以及通过核膜上的 NPC 将 mRNA 和 smallRNA 运输到细胞质等过程在真核生物基因调控中非常重要。NPC 由若干个 NUP 组成。拟南芥 AtNUP160 在低温驯化中有着重要的作用。AtNUP160 在所有组织中均表达且不受低温逆境调节。拟南芥 AtNUP160 - 1 突变体植株对冷害和冻害非常敏感。尽管 AtNUP160 - 1 突变体在常温和低温环境下阻碍了 mRNA 的运输,但是突变体植株细胞核中 mRNA 的积累在冷胁迫下更高。由 AtNUP160 - 1 突变体得知 NPC 的功能与抗冷性之间存在联系。

RNA 解旋酶 DEAD - box 家族参与 RNA 代谢,例如转录、RNA 编辑、RNA 降解和核浆转运。其参与 mRNA 运输和非生物胁迫响应解旋酶的角色通过拟南芥 LOS4 突变体的分析被揭示。感冷 LOS4 - 1 突变体植株在低温驯化过程中降低了 P_{RD29A}::LUC 和 CBF3 的表达,同时推迟了 CBF1 和 CBF2 的表达。LOS4 编码一个 DEAD - box RNA 解旋酶。LOS4 - 2 突变体低温下强烈诱导 CBF2 的表达,从而提高了植株的抗冷性。常温和低温情况下,感冷 LOS4 - 1 突变体植株中 mRNA 运输显著降低。相反,抗冷但感热 LOS4 - 2 突变体植株在冷胁迫下表现正常的 mRNA 运输,但是在常温下 mRNA 在核内的运输存在缺陷。在豌豆中,PDH45 和 PDH47 基因受非生物胁迫上调表达,包括低温处理和 ABA 处理。过表达 PDH45 转基因烟草的抗盐性有所提高。这些研究表明 DEAD - box RNA 解旋酶对 mRNA 转运非常关键。其他解旋酶的功能对植物抗冷和其他非生物胁迫非常重要。核质转运包括 NUP 核质穿梭、转运受体和核运输因子等因素。转运受体在 Ran GTPase 依赖性核质转运中扮演重要的角色。通过 P_{RD29A}::LUC 遗传学筛查,一个输入蛋白 SAD2 被识别,其参与冷胁迫、渗透胁迫和 ABA 处理下的核质转运。SAD2 - 1 无义突变植株能在各种非生物胁迫下增加 P_{RD29A}::LUC 等逆境响应基因的表达。

RNA 聚合酶 Ⅱ 的转录与前体 mRNA 的加工相协调,其对功能性 mRNA 的形成与 mRNA 的转运来说是必要的。聚合酶 Ⅱ CTD 的磷酸化状态在这些过程中具有重要的作用。聚合酶 Ⅱ CTD 磷酸酶通过对聚合酶 Ⅱ 的去磷酸化来控制前体 mRNA 的转录。拟南芥 FRY2/CPL1 是一个 CTD 磷酸酶,其影响 COR 基因的表达。事实上,在低温和 ABA 处理下,FRY2 突变体植株中 CBF 和 COR 基因上调表达。尽管 CBF 基因的表达不受影响,但 FRY2 突变体对冻害也非常敏感,所以 FRY2 在低温驯化过程中主要是通过不依赖于 CBF 的代谢途径来发挥作用的。

1.9.4.2 miRNA 在植物非生物胁迫中的作用

miRNA 和 siRNA 是长度为 21~24 个核苷酸的非编码 RNA,在动物和植物中普遍存在,可抑制基因的表达。miRNA 和 siRNA 通过互补靶 mRNA 指导靶基因断裂或翻译抑制,或是通过诱导靶基因的转录沉默来调控基因表达。非生物胁迫诱导的 miRNA 和 siRNA 下调靶基因的表达,这些靶基因可能作为逆境胁迫的负调控者或决定者;然而其下调表达可能导致靶基因的积累,这些靶基因可能作为抗逆的正调控者或决定者。芯片分析表明 17% 的冷上调基因编码转录因子,而仅仅 7% 的冷下调基因编码转录调控者。因为 miRNA 在非逆境环境中调控各种发育过程,所以冷胁迫下,植物可能使用相同的 miRNA 调控生长和发育。芯片分析表明许多 miRNA 的靶基因参与拟南芥的生长发育调控,同时也响应冷胁迫。冷、ABA、干旱和盐胁迫上调 miR393、miR397b 和 miR402 的表达;冷胁迫可能特异性上调 miR319c 的表达;相反,miR389a.1 受冷、ABA、干旱和盐胁迫下调表达。冷胁迫下 miR393 上调表达,导致其靶基因下调表达。冷诱导 miR393 可能切割泛素化连接酶 E3 的 mRNA,结果导致更少泛素化连接酶 E3 靶蛋白(推测为耐冷正调控因子)的水解。冷上调 miR393 的一个靶基因编码 F-box 蛋白,与酵母 GRR1 相似。由于糖代谢受各种非生物胁迫的影响,植物能利用糖作为信号分子调节非生物胁迫下的生长与发育。冷上调 miR393 可能导致 At4g03190 mRNA 的降解,因为低温处理降低了 At4g03190 的转录水平,因此 miR393 可能融合糖信号响应逆境胁迫。许多不利环境条件(包括低温)能够诱导植物发生渗透胁迫。SOD 作为防御反应的第一道防线能够清除 O_2^-。拟南芥中,渗透胁迫下调 miR398 的表达,导致其靶基因 CSD1 和 CSD2 转录水平上升。因为植物中 miR398 切割 CSD mRNA 靶基因位点是保守的,所以 miR398 可能在渗透胁迫中普遍存在。

1.9.5 翻译后修饰

拟南芥 HOS1 基因突变体在低温条件下可引起 CBF 和其下游靶基因的超诱导。HOS1 编码一个锌指泛素连接酶 E3，该酶在泛素－26S 蛋白酶体代谢中为蛋白质水解提供特异性底物。因此，HOS1 可能作为上游信号分子或者 CBF 转录的负调控者。ICE1 被识别作为 HOS1 的靶蛋白。HOS1 和 ICE1 相互作用，同时在体内外调控 ICE1 的泛素化。低温处理 12 h 后，ICE1 多泛素化和蛋白质水解仅在野生型中被发现，在突变体中不存在。过表达 HOS1 基因能够稳定降低 GFP－ICE1 的蛋白质水平以及 CBF 和 CBF 调控基因的转录水平，导致植株对冻害非常敏感。这些研究结果表明 HOS1 可泛素化 ICE1，负调控 ICE1 靶基因的表达。

SUMO 参与翻译后的蛋白质修饰。SUMO 是一种小泛素相关修饰物。去泛素化是通过 SUMO 蛋白酶从靶基因蛋白中去除 SUMO。蛋白质泛素化/去泛素化在植物响应非生物胁迫、生物胁迫、ABA 和水杨酸信号中扮演重要的角色。SIZ1 无义突变植株对冷和冻害非常敏感。SIZ1 突变体显著降低了 CBF 和其下游靶基因（COR15A、COR47、KIN1）的冷诱导，但是增加了 AtMYB15 的冷诱导。相比 HOS1 促进 ICE1 的降解，SIZ1 在低温驯化中调节 SUMO 结合到 ICE1 的 K393 位置上，进而降低了 ICE1 的多泛素化。SIZ1 调控的 SUMO 可被 ICE1 上 K393R 替换阻止。过表达 ICE1 的转基因拟南芥植株，能够增加 CBF 的冷诱导和提高抗冻性。相反，ICE1 突变体植株和 ICE1 (K393R) 过表达植株能够在冷胁迫下提高 MYB15 的表达，植株表现出对冻害非常敏感。这些结果表明低温驯化过程中，SIZ1 介导的 SUMO 可能维持 ICE1 的稳定性和活性。

1.10 本研究的目的和意义

东北地区位于世界"三大玉米黄金带"上，昼夜温差大，非常适合玉米生长。然而由于地理条件等因素，玉米苗期阶段经常遭受低温侵害，造成产量下降。随着对玉米需求量的不断扩大，提高玉米产量成为当务之急。因此，选育抗冷玉米新种质、新品种，研究玉米冷响应机制就显得尤为重要。

本书以实验室选育的抗冷玉米自交系 W9816 为试验材料,选取三叶期的叶片和根部组织,利用 cDNA – AFLP 的方法比较低温处理前后 DEG 的情况。确定冷响应相关候选基因,利用 RACE – PCR 方法克隆基因全长,并对基因表达进行分析。构建植物表达载体,转化拟南芥对其进行功能分析,阐述候选基因提高转基因植株抗冷性的分子机制,对今后利用冷响应基因提高玉米抗冷性的研究具有重要的理论和实践意义。

第 2 章

冷胁迫下玉米叶片和根部基因表达谱的研究

冷胁迫是重要的非生物胁迫之一,影响植物生长、发育、空间分布和作物产量。为了生存,植物在分子、细胞、生理和生化水平抵抗和适应逆境条件,如植物处于逆境条件下,可通过激发信号转导事件重建细胞稳态。许多逆境响应基因在其他植物中均有报道,它们不但参与植物抗逆(如脱毒酶、抗冻蛋白),同时也参与基因表达调控(如丝裂原活化蛋白激酶)。

植物在非冰冻的低温环境下生长一段时间,能够增强其抗冻能力,从而能够耐受随即发生的冰冻温度,这个适应过程称为低温驯化。大多数温带植物(如冬小麦、燕麦及大麦)通过低温驯化能够耐受组织结冰。许多重要的作物(如水稻、玉米)对低温比较敏感,低温驯化可以降低其对冷害的低温阈值。低温信号能够被各种感受器感知。低温逆境使钙离子通道和质外体空间瞬时诱导钙离子流入细胞质。膜定位类受体蛋白激酶在感受环境信号和激活下游代谢途径中扮演着重要的角色。ICE1-CBF-COR 级联反应是植物响应低温的一条主要信号代谢途径。ICE1 在低温下能够激活 *CBF*3 的转录。CBF 或者 DREB 能够通过识别和结合 *COR* 基因启动子区 DRE/CRT 顺式作用元件来调控 *COR* 基因的表达。3 个相关的钙调蛋白结合转录因子不仅能正调控 *CBF*,同时也能调节 CBF 代谢途径之外的许多早期冷调控基因。有研究人员识别了 5 个早期转录因子(HSFC1、ZAT12、ZF、ZAT10、CZF),它们在非冷处理条件下能够调节 *COR* 基因的表达。尽管如此,仅仅一小部分 *COR* 基因受 CBF 和这 5 个早期转录因子调控,剩余的 *COR* 基因如何被调控仍然未知。

玉米不仅是世界上重要的粮食作物之一,同时在动物饲料和工业产品中发挥关键作用。气候变化往往在许多领域影响粮食生产的能力。在这种条件下,通过常规育种与分子辅助的方法选育耐受逆境的玉米品种迫在眉睫,同时识别新的逆境响应基因和揭示逆境响应机制也具有重要的理论和现实意义。许多关于低温逆境响应的研究在模式植物拟南芥和水稻中被报道,关于玉米冷响应的报道很少。

cDNA-AFLP 是一种用于检测和分离 DEG 的技术,许多逆境诱导基因已被成功分离。本章利用 cDNA-AFLP 技术比较玉米叶片和根部冷胁迫前后基因的表达变化,同时筛选冷响应相关基因。

2.1 试验材料与方法

2.1.1 菌种与载体

菌种为大肠杆菌 Top10,载体为 pMD™18-T 克隆载体(图 2-1)。

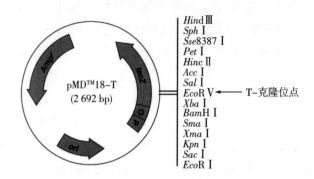

图 2-1　pMD™18-T 克隆载体图谱

2.1.2 主要试剂

RNA 提取试剂(RNAiso Plus)、M-MLV 逆转录酶(RNase H⁻,200 U/μL)、dNTP(2.5 mmol/L)、核糖核酸酶抑制剂(40 U/μL)、Oligo(dT)$_{18}$引物(50 μmol/L)、RNase H(60 U/μL)、T$_4$ DNA 连接酶(350 U/μL)、DNA 聚合酶Ⅰ(4 U/μL)、10×Tango™ buffer、DNase Ⅰ(10 U/μL)、EcoR Ⅰ、Mse Ⅰ、Pst Ⅰ、DL-2000 marker、λ DNA/Hind Ⅲ marker、PrimeScript™ RT Master Mix、SYBR Premix Ex Taq Ⅱ(Tli RNase H Plus)、琼脂糖回收试剂盒、pBR322 DNA/Msp Ⅰ marker、氯化钙、Amp、琼脂糖、cDNA-AFLP 接头、预扩增引物、选择性扩增引物、RNaseA、TEMED、亲和硅烷、剥离硅烷、EDTA、Tris、APS、ddH$_2$O、三氯甲烷、硼酸、异丙醇、丙烯酰胺、甲叉双丙烯酰胺、乙酸、甲酰胺、氢氧化钠、尿素、无水碳酸钠、无水乙醇、氯化钠、柠檬酸三钠、乙醇、三氯甲烷、DEPC、SDS、溴酚蓝、二甲苯青、蛋白胨、酵母提取物以及琼脂粉。

2.1.3 试剂及培养基的配制

2.1.3.1　40%胶贮存液

甲叉双丙烯酰胺 20 g,丙烯酰胺 380 g,定容至 1 L。

2.1.3.2　10×TBE

Tris 108 g,硼酸 55 g,0.5 mol/L EDTA(pH 值为 8.0) 20 mL,定容至 1 L。

2.1.3.3　5%胶工作液

40%胶贮存液 125 mL,尿素 420.24 g,10 × TBE 100 mL,搅拌均匀后定容至 1 L,过滤。

2.1.3.4　变性聚丙烯酰胺凝胶电泳(SDS – PAGE)上样缓冲液

98%甲酰胺 49 mL,10 mmol/L EDTA(pH 值为 8.0) 1 mL,0.25%溴酚蓝 0.125 g,0.25%二甲苯青 0.125 g。

2.1.3.5　10% APS

2 g APS 溶于 18 mL ddH$_2$O,分装, –20 ℃保存。

2.1.3.6　亲和硅烷工作液

在 1 mL 无水乙醇/乙酸(体积比 19∶1)中加入 10 μL 亲和硅烷原液,混合均匀后,用于涂长板。

2.1.3.7　10% SDS

称取 10 g SDS 溶于 100 mL 水中。

2.1.3.8　10 mol/L 氢氧化钠

称取 400 g 氢氧化钠加入 900 mL 水中,完全溶解后定容至 1 L。

2.1.3.9 RNaseA

用 0.15 mol/L 氯化钠和 0.015 mol/L 柠檬酸三钠配成 10 mg/mL 溶液，100 ℃沸水浴处理 15 min 以除去 DNase 活性。

2.1.3.10 LB 培养基

蛋白胨(10 g/L)，酵母提取物(5 g/L)，氯化钠(10 g/L)，调 pH 值至 7.2。

2.1.3.11 Amp(100 mg/mL)

取 1 g Amp 溶于 10 mL ddH$_2$O 中，过滤灭菌，分装，-20 ℃保存。

2.1.3.12 50×TAE 缓冲液

Tris 121 g，乙酸 28.55 mL，0.5 mol/L EDTA(pH 值为 8.0)50 mL，定容至 500 mL。

2.1.4 主要仪器

PCR 扩增仪、高速低温离心机、高速离心机、电泳仪、水平电泳槽、垂直板电泳槽、恒温水浴锅、凝胶成像系统、恒温培养箱、水平摇床、可控温摇床以及 real-time PCR 系统等。

2.1.5 植物材料

以实验室筛选的玉米抗冷自交系 W9816 为试验材料，使用 75% 乙醇对种子表面消毒 3 min，随后纯化水漂洗 3 次，置于两层湿润滤纸之间萌发，25 ℃暗培养 3 d。将均匀萌发的种子转移到含有草炭土、蛭石和珍珠岩(体积比为 10:1:1)移栽基质的小盆中，将其放入人工气候箱中培养两周，温度为 25 ℃/20 ℃(白天/夜晚)，光周期为 14 h/10 h(白天/夜晚)，相对湿度为 90%/95%(白天/夜晚)，光量子通量密度为 450 μmol·m^{-2}·s^{-1}。

2.1.6 冷胁迫处理

待玉米植株生长到三叶期时(两周后)，低温(4 ℃)光照下处理 12 h，相对湿度为90%/95%(白天/夜晚),光量子通量密度为450 $\mu mol \cdot m^{-2} \cdot s^{-1}$。在正常温度(25 ℃)下生长的植株作为对照,光照条件等与冷处理相同。同时分别采集10株冷胁迫处理和对照的叶片及根部组织,分别混合后放入液氮内速冻,-80 ℃储存于冰箱待用。

2.1.7 总RNA的提取

利用RNAiso Plus提取玉米叶片和根部组织的总RNA。

(1)取冷胁迫(4 ℃)处理12 h和对照(25 ℃)相同部位叶片及根部组织,迅速转移到液氮预冷的研钵中,立即加入1 mL RNAiso Plus,充分匀浆。

(2)将匀浆液转移至离心管中,室温(15~30 ℃)静置5 min。

(3)向匀浆液中加入200 μL三氯甲烷,盖紧离心管盖子,混合至溶液乳化呈乳白色。

(4)室温静置5 min。

(5)12 000 ×g、4 ℃离心15 min。

(6)吸取上清液移入新的离心管中,切勿吸出白色中间层。

(7)向上清液中加入等体积的异丙醇,上下颠倒离心管充分混匀后,室温下静置10 min。

(8)12 000 ×g、4 ℃离心10 min,离心后试管底部出现RNA沉淀。

(9)小心弃去上清液,切勿触及沉淀,加入与上清液等量的75%乙醇。轻轻上下颠倒洗涤离心管管壁,7 500 ×g、4 ℃离心5 min后小心弃去上清液,切勿触及沉淀。

(10)打开离心管盖子,室温干燥沉淀几分钟。沉淀干燥后,加入20 μL的DEPC水溶解沉淀。不可以离心或加热干燥,否则RNA将会很难溶解。

(11)取1 μL总RNA,稀释200倍,分光光度计测OD值。OD_{260}/OD_{280}值在1.8~2.0范围内为合格。取2 μL RNA用1%琼脂糖凝胶电泳进行检测。

2.1.8 基因组DNA的去除

(1)使用DNase I 除去RNA混有的基因组DNA,其反应混合液组成见表2-1。

表 2-1 去除 DNA 反应混合液组成

成分	用量
总 RNA	20 μg
10×DNase Ⅰ	5 μL
DNase Ⅰ(RNase - free)	2 μL
核糖核酸酶抑制剂(40 U/μL)	0.5 μL
RNase - free ddH$_2$O	补足到 50 μL

(2)37 ℃反应 20~30 min。

(3)加入 0.5 mol/L EDTA 2.5 μL,混匀,80 ℃加热处理 2 min。

(4)用 DEPC 水定容至 100 μL。

(5)加入 10 μL 3 mol/L 乙酸钠和 250 μL 冷乙醇,混匀后 -80 ℃放置 20 min。

(6)4 ℃、12 000 r/min 离心 10 min,弃上清液。

(7)加入 70% 冷乙醇洗净,4 ℃、12 000 r/min 离心 5 min,弃上清液。

(8)干燥沉淀。

(9)用适量的 DEPC 水溶解后,进行电泳,确认是否除去基因组 DNA,同时测定 RNA 浓度。

2.1.9 双链 cDNA 的合成

2.1.9.1 cDNA 第一链的合成

cDNA 第一链合成混合液成分见表 2-2。

表 2-2　cDNA 第一链合成混合液组成

成分	用量
总 RNA	1 μg
5×M-MLV	4 μL
Oligo (dT)$_{18}$ 引物(50 μmol/L)	2 μL
dNTP(2.5 mmol/L)	4 μL
M-MLV 逆转录酶(200 U/μL)	1 μL
核糖核酸酶抑制剂(40 U/μL)	0.5 μL
RNase-free dH$_2$O	补足到 20 μL

42 ℃ 1 h,70 ℃ 15 min,冰上冷却,-20 ℃ 保存。取 5 μL cDNA 第一链产物用 1%琼脂糖凝胶电泳检测。

2.1.9.2　cDNA 第二链的合成

cDNA 第二链合成混合液成分见表 2-3。

表 2-3　cDNA 第二链合成混合液组成

成分	用量/μL
cDNA 第一链	10
10×T$_4$DNA 连接酶 buffer	3
10×DNA 聚合物 I buffer	3
RNase H(60 U/μL)	0.5
T$_4$DNA 连接酶(350 U/μL)	0.5
dNTP(2.5 mmol/L)	6
DNA 聚合酶 I (4 U/μL)	2
ddH$_2$O	补足到 30 μL

16 ℃ 2.5 h,80 ℃ 10 min,冰上急冷 2 min,-20 ℃ 保存。取 5 μL 双链 cDNA 用 1%琼脂糖凝胶电泳检测。

2.1.10 cDNA-AFLP 程序

2.1.10.1 酶切

使用 *Mse* I/*Pst* I 和 *Mse* I/*EcoR* I 两种限制性内切酶组合对双链cDNA进行酶切。酶切体系见表2-4。

表2-4 酶切体系组成(30 μL)

成分	用量/μL
10 × Tango^{Tm} buffer	6
双链 cDNA	22.5
Mse I	0.75
EcoR I (*Pst* I)	0.75

37 ℃ 3 h,65 ℃ 7.5 h,80 ℃ 20 min,冰上急冷 2 min,-20 ℃保存。取 5 μL酶切产物用1%琼脂糖凝胶电泳检测。

2.1.10.2 接头的连接

EcoR I 接头序列:5'- CTCGTAGACTGCGTACC -3';3'- CATCTGACGCATGGTTAA -5'。

Pst I 接头序列:5'- CTCGTAGACTGCGTACATGCA -3';3'- CATCTGACGCATGT -5'。

Mse I 接头序列:5'- GACGATGAGTCCTGAG -3';3'- TACTCAGGACTCAT -5'。

接头连接体系见表2-5。

表2-5 接头连接体系组成(40 μL)

成分	用量/μL
酶切产物	20
EcoR I (Pst I)接头(5 μmol/L)	1
Mse I 接头(50 μmol/L)	1
T_4 DNA 连接酶(350 U/μL)	0.5
10 × T_4 DNA 连接酶 buffer	4
ddH_2O	13.5

16 ℃过夜连接。

2.1.10.3 预扩增

连接产物稀释10倍,进行预扩增。引物与体系见表2-6、表2-7。

表2-6 预扩增引物

引物	序列(5'—3')
Mse I 预扩增引物	GATGAGTCCTGAGTAA
Pst I 预扩增引物	GACTGCGTACATGCAG
EcoR I 预扩增引物	GACTGCGTACCAATTC

表2-7 预扩增体系(20 μL)

成分	用量/μL
模板	5
dNTP(2.5 mmol/L)	1.6
10 × PCR buffer	2
Mse I 预扩增引物(10 μmol/L)	1
EcoR I (Pst I)预扩增引物(10 μmol/L)	1
rTaq(5 U/μL)	0.2
ddH_2O	9.2

PCR 反应程序:94 ℃ 变性 30 s,56 ℃ 退火 1 min,72 ℃ 延伸 1 min,反应进行 25 个循环。取 5 μL PCR 产物用 1% 琼脂糖凝胶电泳检测。

2.1.10.4　选择性扩增

选择性扩增引物见表 2-8。

表2-8 选择性扩增引物

EcoR I 引物(8)	Mse I 引物(10)	Pst I 引物(16)
E1:GACTGCGTACCAATTCAAC	M1: GATGAGTCCTGAGTAACAA	P1: GACTGCGTACATGCAGAA
E2:GACTGCGTACCAATTCAAG	M2: GATGAGTCCTGAGTAACAC	P2: GACTGCGTACATGCAGAC
E3:GACTGCGTACCAATTCACA	M3: GATGAGTCCTGAGTAACAG	P3: GACTGCGTACATGCAGAG
E4:GACTGCGTACCAATTCACT	M4: GATGAGTCCTGAGTAACAT	P4: GACTGCGTACATGCAGAT
E5:GACTGCGTACCAATTCACC	M5: GATGAGTCCTGAGTAACTA	P5: GACTGCGTACATGCAGCA
E6:GACTGCGTACCAATTCACG	M6: GATGAGTCCTGAGTAACTC	P6: GACTGCGTACATGCAGCC
E7:GACTGCGTACCAATTCAGC	M7: GATGAGTCCTGAGTAACTG	P7: GACTGCGTACATGCAGCG
E8:GACTGCGTACCAATTCAGG	M8: GATGAGTCCTGAGTAACTT	P8: GACTGCGTACATGCAGCT
	M9: GATGAGTCCTGAGTAAGGA	P9: GACTGCGTACATGCAGGA
	M10:GATGAGTCCTGAGTAAGGC	P10:GACTGCGTACATGCAGGC
		P11:GACTGCGTACATGCAGGT
		P12:GACTGCGTACATGCAGGT
		P13:GACTGCGTACATGCAGTA
		P14:GACTGCGTACATGCAGTC
		P15:GACTGCGTACATGCAGTG
		P16:GACTGCGTACATGCAGTT

预扩增产物稀释 15 倍,取 5 μL 为模板用于选择性扩增,反应体系见表 2-9。

表 2-9 选择性扩增反应体系(20 μL)

成分	用量/μL
模板	5
dNTP(2.5 mmol/L)	1.6
10×PCR buffer	2
EcoR I (Pst I)选择性扩增引物(10 μmol/L)	1
Mse I 选择性扩增引物(10 μmol/L)	1
rTaq(5 U/μL)	0.2
ddH$_2$O	9.2

选择性扩增用 Touch down PCR 进行,程序如下:

94 ℃ 2 min

94 ℃ 30 s

65 ℃ 1 min(每循环降 0.7 ℃) ⎫

72 ℃ 1 min ⎭ 12 个循环

94 ℃ 30 s

56 ℃ 1 min ⎫ 23 个循环

72 ℃ 1 min ⎭

4 ℃终延伸

2.1.10.5 SDS-PAGE

(1)玻璃板清洗:用水将玻璃板彻底清洗干净,然后用纯化水冲洗一遍,干燥后,用无水乙醇擦拭一遍,室温干燥。

(2)玻璃板的处理:将长板用亲和硅烷工作液擦拭两遍,室温干燥备用。短板涂上 500 μL 剥离硅烷,擦拭两遍,室温干燥备用。注意不要污染,否则容易造成粘板。

(3)板的组装:先将长板放在水平台上,涂亲和硅烷面向上;左右两侧放上边条,然后将短板涂剥离硅烷的面向下,与长板对齐,每边用三四个夹子夹紧。

(4) 变性测序胶的配制:向 5% 胶工作液 90 mL 中加入 10% APS 320 μL、TEMED 80 μL。

(5) 灌胶:将胶充分混匀后,沿灌胶口轻轻灌进玻璃板,轻轻拍打玻璃板,防止气泡出现。待胶流动到底部,在灌胶口轻轻插入梳子,在梳子处夹上夹子,室温静置 2 h 以上,使其充分聚合(若胶过夜聚合,在胶的两头封上保鲜膜以防胶干或流出造成气泡)。

(6) 电泳:取下含有聚合胶的玻璃板上的梳子,用大量水冲洗玻璃板,使其表面没有胶残留,之后用纯化水冲洗干净。将玻璃板装入电泳槽,夹紧。下槽电极缓冲液为 $1 \times TBE$,上槽电极缓冲液为 $0.5 \times TBE$。检查设备连接正常后,先在功率为 90 W 下预电泳 30 min。在预电泳时,在每管选择性扩增产物中加入 3 μL 上样缓冲液,94 ℃变性 10 min,冰上急冷。预电泳结束后,用 1 mL 枪轻轻吹打胶面,除去尿素和碎胶等,插入梳子,梳子齿插入胶面 1~2 mm,轻轻用枪吹打每个点样孔,每个点样孔中加入 6 μL 变性后的选择性扩增产物(注意样品要一直在冰上,防止复性;每个点样孔缓慢加入样品,点样量尽量一致),在 60 W 恒功率下电泳,直到溴酚蓝到达板底部(大约 3.5 h)为止。

(7) 卸板:电泳结束后,小心取下玻璃板,取下短板,去掉梳子和边条,胶应该粘在长玻璃板上。

2.1.10.6　DNA 银染

(1) 脱色:取合适的塑料盒,倒入 2 L 新配制的固定液(10% 乙酸),将带胶的长玻璃板轻轻放入,在摇床上轻轻摇动至全部脱色(大约 30 min)。

(2) 冲洗:用纯化水冲洗胶板 2~3 次,每次 2 min。

(3) 染色:将板放入染色液(每 2 L 纯化水中加入 2 g 硝酸银和 3 mL 甲醛溶液)中,轻轻摇动 30 min。

(4) 冲洗:用纯化水冲洗胶板不超过 5 s。

(5) 显影:将冲洗过的胶板迅速转移到 2 L 预冷的显影液(2 L 纯化水中加入 60 g 无水碳酸钠,预冷;使用前需加入 3 mL 甲醛溶液和 400 μL

10 mg/mL 硫代硫酸钠溶液)中,轻轻摇动,直至带纹出现。

(6)定影:将板放入固定液中轻摇 3~5 min。

(7)漂洗:用纯化水漂洗胶板两次,每次 2~3 min。室温下自然干燥,拍照,也可以永久保存。

2.1.10.7 目的条带的回收

(1)用锋利的手术刀将胶上的目的带切割后,放入 1.5 mL 离心管中,加入 200 μL ddH$_2$O 冲洗两遍。

(2)加入 30 μL ddH$_2$O,用枪尖捣碎,37 ℃ 温浴 8 h,沸水浴 15 min。

(3)12 000 ×g 离心 2 min。

(4)吸上清液到灭菌的 0.5 mL 离心管中,−20 ℃ 保存备用。

2.1.11 再次 PCR

以回收产物 5 μL 为模板,利用相同的选择性扩增引物与 PCR 条件重新扩增。PCR 体系和条件见 2.1.10.4。

2.1.12 目的片段的琼脂糖回收

用普通琼脂糖凝胶 DNA 回收试剂盒回收目的片段,方法如下:

(1)向吸附柱 CA2(吸附柱放入收集管中)中加入 500 μL 平衡液 BL(由试剂盒提供),12 000 r/min(约 13 400 ×g)离心 1 min,倒掉收集管中的废液,将吸附柱重新放回收集管中。

(2)将单一的目的 DNA 条带从琼脂糖凝胶中切下(尽量切除多余部分),放入干净的离心管中,称重。

(3)向胶块中加入等体积 PN 溶液(如果凝胶质量为 0.1 g,其体积可视为 100 μL,则加入 100 μL PN 溶液,PN 溶液由试剂盒提供),50 ℃ 水浴放置,其间不断温和地上下翻转离心管,以确保胶块充分溶解。

(4)将上一步所得溶液加入吸附柱 CA2(吸附柱放入收集管中)中,室温放置 2 min,12 000 r/min(约 13 400 ×g)离心 30~60 s,倒掉收集管中的废液,将吸附柱 CA2 放入收集管中。

(5)向吸附柱 CA2 中加入 600 μL 漂洗液 PW(由试剂盒提供,使用前请先检查是否已加入无水乙醇),12 000 r/min(约 13 400×g)离心 30~60 s,倒掉收集管中的废液。

(6)重复步骤(5)。

(7)将吸附柱 CA2 放回收集管中,12 000 r/min(约 13 400×g)离心 2 min,尽量除尽漂洗液。将吸附柱 CA2 于室温放置数分钟,彻底晾干,以防止残留的漂洗液影响下一步试验。

(8)将吸附柱 CA2 放入一个干净的离心管中,向吸附膜中间位置悬空滴加适量洗脱缓冲液 EB(由试剂盒提供),室温放置 2 min。12 000 r/min(约 13 400×g)离心 2 min,收集 DNA 溶液。

2.1.13 测序

2.1.13.1 目的片段与克隆载体连接

连接体系见表 2-10。

表 2-10 连接体系(10 μL)

成分	用量/μL
pMD^{TM} 18-T	1
Solution Ⅰ	5
回收片段	2.5
ddH$_2$O	1.5

16 ℃反应 30 min。

2.1.13.2 大肠杆菌 Top10 感受态细胞的制备(氯化钙法)

(1)将在-80 ℃冰箱中保存的菌种取出进行划线培养,从大肠杆菌 Top10 平板上挑取一个单菌落,接种于 3 mL LB 液体培养基中,37 ℃振荡培养过夜。

(2)取 1 mL 菌液加入含有 50 mL LB 液体培养基的三角瓶中,37 ℃恒温,

180 r/min 振荡培养 2~3 h。

(3)提前预冷 50 mL 离心管,当菌液 OD_{600} 达到 0.4 左右时,将菌液转入预冷离心管,冰上放置 10 min。

(4)4 ℃、8 000 r/min 离心 10 min,收集菌体。

(5)倒掉培养液,将管倒置于灭菌的滤纸上,使管内培养液流尽。

(6)用 0.1 mol/L 冰预冷氯化钙 10 mL 悬浮沉淀,吹打混匀,立即放在冰上 30 min。

(7)4 ℃、8 000 r/min 离心 10 min,收集菌体。

(8)弃培养液,加入冰预冷的含有 15% 甘油的 0.1 mol/L 氯化钙 2 mL 悬浮细胞,冰上放置。

(9)每管 100 μL 分装,置于液氮中速冻,然后于 -80 ℃ 保存。

2.1.13.3 重组质粒的转化

(1)在超净工作台上,将 10 μL 的重组子加入 Top10 感受态细胞里,用枪吹打混匀后,置于冰中 30 min。

(2)42 ℃ 热激 90 s。

(3)放回冰中 3~5 min。

(4)在超净工作台上,加入 700 μL 不含 Amp 的 LB 液体培养基,用枪吹打混匀。

(5)在恒温振荡摇床中,37 ℃、180 r/min 振荡 1 h。

(6)将菌液于 8 000 r/min 离心 2 min 后,弃上清液。

(7)用枪吹打悬浮菌液,将 50 μL 菌液均匀涂到含 Amp 的 LB 平板上,放入恒温培养箱中 37 ℃ 培养 12~16 h。

(8)挑取单菌落放入 1.5 mL 的离心管中,加入 1 mL 含 Amp 的 LB 液体培养基,于恒温振荡摇床 37 ℃、180 r/min 振荡培养 4 h。

2.1.13.4 重组质粒的 PCR 鉴定

PCR 鉴定体系见表 2-11。

第 2 章　冷胁迫下玉米叶片和根部基因表达谱的研究

表 2-11　PCR 鉴定体系(50 μL)

成分	用量/μL
菌液	1
10×PCR buffer	5
dNTP(各 2.5 mmol/L)	4
PCR 引物序列(F)(10 μmol/L)	0.5
PCR 引物序列(R)(10 μmol/L)	0.5
rTaq(5 U/μL)	0.25
ddH_2O	38.75

PCR 反应程序：

94 ℃ 5 min

94 ℃ 30 s ⎫
55 ℃ 30 s ⎬ 30 个循环
72 ℃ 30 s ⎭

4 ℃ 终延伸

取 5 μL PCR 产物进行 1% 琼脂糖凝胶电泳检测。

2.1.13.5　PCR 扩增条带的测序

将 PCR 阳性菌液送样测序。

2.1.14　序列分析

测序结果去掉载体序列后在 NCBI BLASTX 和 MaizeGDB 数据库进行比对，寻找其同源序列。通过 NCBI BLASTX 和 MaizeGDB 搜索的已知基因编码蛋白质的功能进行基因功能的分类。

2.1.15　real-time PCR

利用 ABI 7500 real-time PCR 系统进行 real-time PCR 分析。选择 16 个冷响应相关基因，引物见表 2-12。RNA 的提取见 2.1.7，反转录使用 PrimeScript™ RT Master Mix。cDNA 稀释 5 倍，按照试剂盒 SYBR Premix Ex *Taq* Ⅱ(Tli RNase H Plus)说明进行 real-time PCR。引物设计在 100~300 bp 之间，且不能出现引物二聚体和非特异性扩增。利用 $2^{-\Delta\Delta Ct}$ 方法确定各基因的相对定量情况。

表 2-12　候选基因和内参特异引物

名称	real-time PCR 引物序列(F,5'—3')	real-time PCR 引物序列(R,5'—3')	T_m/℃
T31	TCTTGGAGGCAGTTGTGA	ACCTGTTGACTTTCTTCAGAGG	55
T49	TGCGACGCAATGAACAATAT	CAATAACAAGGTAACGCCAAA	55
T203	ATGCTTCGGAGAGTGGTCT	TGTCATTGTGTAAACGGGTC	55
T11	AGATAAGCTACGACCTACGAGG	GCAGGGATTTTACAGAGAACAC	55
T163	TCTCCATCGTGCTCTTCAA	ACTTCACCAATCGGTTCTTC	55
T147	GTTGGCTCTTTGCTGTCTC	CCTGAACGCTGAACCTACA	55
T194	ATGGCGACATTATCAACATC	ATGCACTCTTCAGTGACCTT	55
T258	AAATGGCAAGGTGAAGGC	TCGACACTGGTAGCGAAG	55
T207	ATTTGTTGTTGTCTTTGCGT	GTGTGGGTATCATCTGGTC	55
T242	GGATGGTAAGGAGATAGGCAT	GGCTAAAGGTCCAGTTCCA	55
T199	CAAGATTCTATTGGCACGG	GACCCATTAGCAACATTGG	55
T2	CGTTGATGAAGCAGATGTCC	CTTGTGTGAGAACCGCAAT	55
T162	AAGAGGAAAGAGGGCGAGT	ATGTGTTCGTAGGCTCCGT	55
T8	GTGGCTGTCTGTCATCGGC	TGATCTCAAGGCGCTTCTCA	55
T60	GGCGGCGACAGGCTTCTTCT	AGATCGACGGATGGGTTG	55
T252	GCTGAAAGAGGATGACGCA	TGGGAGTTCCACAGGTTGT	55
*ZmActin*1	CGATTGAGCATGGCATTGTCA	CCCACTAGCGTACAACGAA	55
γ-tubulin	CAATGGGATTCAGAACAGCGA	AGCAATAGAGTGGCACAGGAC	55

cDNA 合成的反应体系见表 2-13。

表 2-13 cDNA 合成反应体系

成分	用量/μL
5 × PrimeScript™ RT Master Mix	2
总 RNA	1
RNase-free dH$_2$O	7

轻柔混合后进行反转录反应,条件如下:

37 ℃ 15 min(反转录反应)

85 ℃ 5 s(反转录酶的失活反应)

4 ℃ 终延伸

real-time PCR 体系见表 2-14。

表 2-14 real-time PCR 体系(20 μL)

成分	用量/μL
SYBR Premix Ex *Taq* Ⅱ (Tli RNase H Plus)	10
PCR 引物序列(F)(10 μmol/L)	0.8
PCR 引物序列(R)(10 μmol/L)	0.8
ROX Reference Dye Ⅱ	0.4
cDNA 溶液	2
ddH$_2$O	6

real-time PCR 条件如下:

95 ℃ 30 s

95 ℃ 5 s ⎫
60 ℃ 34 s ⎭ 40 个循环

95 ℃ 15 s

60 ℃ 1 min

95 ℃ 15 s

2.2 结果与分析

2.2.1 RNA 的提取

提取正常温度(25 ℃)和冷处理(4 ℃)12 h 下玉米叶片和根总 RNA。从图 2-2 可以看出,提取的总 RNA 比较完整,可以清晰地看到 28S RNA、18S RNA 和 5S RNA 3 条条带。紫外分光光度计检测 RNA 的样品浓度为 1 μg/μL,OD_{260}/OD_{280} 值在 1.8～2.0 之间,提取质量较好,可用于进一步反转录试验。

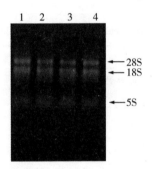

图 2-2 总 RNA 提取结果
1.常温(25 ℃)叶片;2.4 ℃处理 12 h 叶片;
3.常温(25 ℃)根;4.4 ℃处理 12 h 根

2.2.2 cDNA 合成

利用总 RNA 为模板,反转录 cDNA 的第一条链,进而合成 cDNA 第二条链。从图 2-3 可以看出,双链 cDNA 在 2 000 bp 下呈弥散带,证明 cDNA 较为完整,可用于下一步酶切试验。

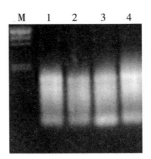

图 2-3 双链 cDNA 合成结果

1. 常温(25 ℃)叶片；2. 4 ℃处理 12 h 叶片；

3. 常温(25 ℃)根；4. 4 ℃处理 12 h 根；

M. λ DNA/*Hind*Ⅲ marker

2.2.3　cDNA-AFLP 分析

2.2.3.1　双链 cDNA 的酶切

利用 *Mse*Ⅰ/*Eco*RⅠ 和 *Mse*Ⅰ/*Pst*Ⅰ 限制性酶切组合对双链 cDNA 进行酶切，结果如图 2-4 所示。2 000 bp 以下呈弥散带，且分布均匀，表明 cDNA 酶切结果完全、充分。

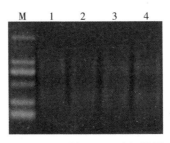

图 2-4 双链 cDNA 酶切结果

1. 常温(25 ℃)叶片；2. 4 ℃处理 12 h 叶片；

3. 常温(25 ℃)根；4. 4 ℃处理 12 h 根；

M. DL-2000 marker

2.2.3.2 预扩增和选择性扩增

双链 cDNA 酶切完全后,末端连接相应的限制性内切酶接头。连接产物稀释 10 倍用于预扩增。预扩增结果显示,扩增带大多为 1 000 bp 以下,适合选择性扩增(图 2-5)。5 μL 的欲扩增产物稀释 15 倍用于选择性扩增。利用 174 对引物组合(包括 72 对 *Mse* I/*Eco*R I 和 102 对 *Mse* I/*Pst* I 引物组合)进行选择性扩增,扩增产物进行 SDS-PAGE 电泳。

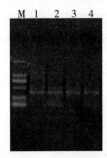

图 2-5 预扩增结果
1. 常温(25 ℃)叶片;2. 4 ℃处理 12 h 叶片;
3. 常温(25 ℃)根;4. 4 ℃处理 12 h 根;
M. DL-2000 marker

2.2.3.3 冷胁迫下玉米叶片和根部 DEG 的分析

cDNA-AFLP 表达谱用于检测冷胁迫后玉米叶片和根部 DEG。通过 cDNA-AFLP 分析,在叶片发现 6 829 个 TDF,在根部发现 6 955 个 TDF(图 2-6)。每个引物组合有 30~50 条 AFLP 扩增带,其大小介于 70~600 bp 之间。尽管大多数条带在冷胁迫后没有明显变化,但在叶片中仍然检测到 620 个 DEG,在根部中检测到 531 个 DEG。这些 DEG 显示出不同的表达模型。

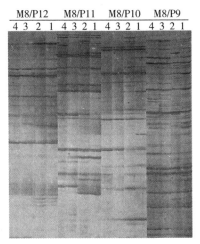

图 2-6　cDNA-AFLP DEG 扩增图

1. 常温(25 ℃)叶片；2. 4 ℃处理 12 h 叶片；

3. 常温(25 ℃)根；4. 4 ℃处理 12 h 根

注：M8/P9、M8/P10、M8/P11、M8/P12 代表了不同的选择性 AFLP 引物。

2.2.3.4　DEG 分类

根据叶片和根部获得的 DEG 情况，可将 DEG 分为以下几类：诱导型(induced)、上调型(up-regulated)、快速抑制型(rapid switched-off)和下调型(down-regulated)(图 2-7)。

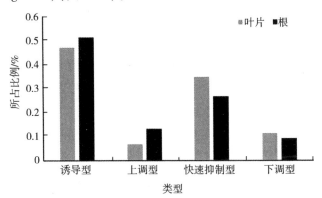

图 2-7　DEG 分类

291 个 DEG(叶片)和 271 个 DEG(根)仅在冷处理后诱导表达(诱导型，Ⅰ类)；42 个 DEG(叶片)和 70 个 DEG(根)在常温下表达量很少，但是在冷处理 12 h 后表达量大幅度增加（上调型，Ⅱ类）；216 个 DEG(叶片)和 141 个 DEG(根)在正常温度下表达，但是冷处理后不表达(快速抑制型，Ⅲ类)；71 个 DEG(叶片)和 49 个 DEG(根)在常温下表达，冷胁迫后表达量下降(下调型，Ⅳ类)。值得注意的是，相比于Ⅱ类和Ⅳ类，Ⅰ类和Ⅲ类在 DEG 中占有较大比例，Ⅰ类中的基因可能作为逆境胁迫下的正调控者或决定者，Ⅲ类中的基因可能作为逆境胁迫下的负调控者或决定者。其中，仅仅 61 个上调的 DEG 和 32 个下调的 DEG 在两种组织中同时具有一致的表达趋势，大多数 DEG 冷胁迫后在两种组织中体现出不同的表达趋势。

2.2.4 差异表达片段的回收测序

采用压碎浸泡法从 PAGE 中共回收了 67 条差异表达片段。利用相同的选择性扩增引物及扩增条件对回收的差异表达片段再次扩增，经琼脂糖回收，最终成功回收 67 个差异表达片段（图 2-8）。

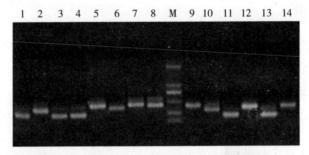

图 2-8 部分差异表达片段的 PCR 扩增结果
1~14. 差异表达片段再次 PCR 结果；M. DL-2000 marker

2.2.5 序列分析

通过 NCBI BLASTX 和 MaizeGDB 数据库分析，根据功能将 DEG 主要分为 4 类，即信号转导(10,15%)、转录调控(9,13%)、翻译和翻译后修饰(10,15%)以及细胞代谢与组织(24,36%)，6 个 DEG 为推测编码蛋白(9%)，8 个 DEG 在数据库中没有匹配(12%)。见表 2-15。

表2-15 DEGs的同源性分析

引物组合	DEG	大小/bp	同源蛋白	物种	GenBank登录号	Chr^c	E值
细胞代谢与组织(24)							
M7/P11	T8	323	S-adenosylmethionine decarboxylase proenzyme（S-腺苷-L-甲硫氨酸脱羧酶酶原）	玉米	gi|413922996	Chr4	1.507×10^{-121}
M8/P9	T49^b	206	carbonic anhydrase（碳酸酐酶）	玉米	gi|414589702	Chr2	4.357×10^{-85}
M8/P11	T60	266	shikimate kinase family protein（莽草酸激酶家族蛋白）	玉米	gi|414868512	Chr1	5.521×10^{-120}
M4/E6	T218	237	cytokinin inducible protease 1（细胞分裂素诱导蛋白酶1）	玉米	gi|413916758	Chr10	6.328×10^{-114}
M4/E1	T208	381	alpha-L-arabinofuranosidase family protein 1（α-L-阿拉伯呋喃糖酶家族蛋白1）	玉米	gi|413920612	Chr4	6.512×10^{-96}
M7/E3	T241	113	D-glycerate-3-kinase（D-甘油酸-3-激酶）	玉米	gi|226491003	Chr6	2.700×10^{-25}
M8/E6	T258^b	350	NADH-dependent glutamate synthase 1（NADH依赖性谷氨酸合成酶1）	玉米	gi|414880974	Chr3	7.853×10^{-80}
M4/E1	T207^b	417	nucleotide/sugar transporter family protein（核苷酸/糖转运体家族蛋白）	玉米	gi|670437331	Chr9	1.952×10^{-111}
M7/P7	T41	209	sulfite oxidase（亚硫酸盐氧化酶）	玉米	gi|414869647	Chr1	8.073×10^{-78}

续表

引物组合	DEG	大小 bp	同源蛋白	物种	GenBank 登录号	Chr[c]	E 值
M9/P3	T163[b]	327	chloride channel E（E 型氯离子通道）	玉米	gi670389456	Chr3	2.094×10^{-60}
M6/P3	T195	242	light-harvesting complex-like protein OHP2（捕光复合体蛋白 OHP2）	玉米	gi413950479	Chr8	5.374×10^{-65}
M9/P1	T141[a]	522	cyclin-related protein（细胞周期相关蛋白）	水稻	gi670361173	Chr1	1.998×10^{-82}
M4/E6	T217	551	peroxisome biosynthesis protein PAS1-like（过氧化物酶体生物合成蛋白 PAS1）	玉米	gi670430762	Chr8	0
M5/E2	T246	126	NADH dehydrogenase Ⅰ subunit 1（NADH 脱氢酶Ⅰ亚基 1）	拟南芥	gi413954178	Chr9	1.190×10^{-33}
M4/P5	T231	434	thioredoxin domain-containing protein 9 homolog（硫氧还蛋白结构域蛋白 9）	玉米	gi413922131	Chr4	1.655×10^{-77}
M8/P12	T61	391	SKIP interacting protein 7（SKIP 相互作用蛋白 7）	水稻	gi413944054	Chr6	8.013×10^{-150}
M7/P5	T31	321	AAA-type ATPase family protein（AAA 型 ATPase 家族蛋白）	水稻	gi414591153	Chr2	1.575×10^{-71}
M9/P2	T159[b]	176	photosystem Ⅰ reaction center subunit psaK（光系统Ⅰ反应中心亚基 psaK）	玉米	gi226508528	Chr7	1.032×10^{-70}
M9/P1	T146	185	SET-domain containing protein family（SET 结构域蛋白家族）	玉米	gi414586230	Chr2	1.079×10^{-80}

续表

引物组合	DEG	大小/bp	同源蛋白	物种	GenBank 登录号	Chr	E 值
M9/P2	T153	379	O-methyltransferase ZRP4-like（O-甲基转移酶 ZRP4）	玉米	gi|670443834	Chr10	2.729×10^{-164}
M7/P2	T22	251	CMO protein（胆碱单加氧酶）	玉米	gi|413943240	Chr6	4.82×10^{-61}
M8/E6	T257	376	EKC/KEOPS complex subunit bud32（EKC/KEOPS 复合体亚基 bud32）	小米	gi|413934180	Chr5	5.186×10^{-131}
M9/P2	T157[b]	259	K^+ efflux antiporter 3（K^+ 逆向转运体 3）	小米	gi|414877582	Chr3	2.026×10^{-84}
M7/P12	T10	478	peroxidase 51（过氧化物酶 51）	小米	gi|414872974	Chr1	4.494×10^{-148}
信号转导(10)							
M6/P6	T199[a]	462	response regulator receiver domain containing protein（应答调节蛋白）	短花药野生稻	gi|414588535	Chr2	9.150×10^{-31}
M7/P11	T6	407	ACT-domain containing protein kinase family protein（ACT 结构域蛋白激酶家族蛋白）	玉米	gi|293331679	Chr2	2.130×10^{-31}
M6/P6	T203	321	calmodulin-binding protein（钙调结合蛋白）	短花药野生稻	gi|413916568	Chr10	1.095×10^{-142}
M9/P3	T162	421	LRR-like protein kinase（LRR 型蛋白激酶）	玉米	gi|670424888	Chr7	9.241×10^{-105}

续表

引物组合	DEG	大小/bp	同源蛋白	物种	GenBank登录号	Chr^c	E 值
M8/E2	T252^b	129	CBL–interacting protein kinase family protein（CBL蛋白激酶）	玉米	gi 414883532	Chr7	1.043×10^{-39}
M7/P6	T242	127	phospholipid scramblase family protein（磷脂爬行酶家族蛋白）	玉米	gi 670434175	Chr8	1.359×10^{-37}
M7/P6	T36	144	tetratricopeptide repeat protein KIAA0103（四肽重复蛋白 KIAA0103）	玉米	gi 413932445	Chr5	1.115×10^{-34}
M3/E1	T287^a	277	Sec14p–like phosphatidylinositol transfer family protein（Sec14类磷脂酰肌醇转运家族蛋白）	玉米	gi 966039092	Chr1	4.432×10^{-126}
M7/P13	T16	79	patatin–like PLA（patatin类磷脂酶A）	玉米	gi 670366127	Chr1	4.083×10^{-26}
M2/P2	T86	409	Tat pathway signal sequence family protein（Tat途径信号序列家族蛋白）	玉米	gi 226530700	Chr8	8.116×10^{-175}

转录调控(9)

引物组合	DEG	大小/bp	同源蛋白	物种	GenBank登录号	Chr^c	E 值
M7/P6	T33	425	mediator of RNA polymerase Ⅱ transcription subunit 15a–like isoform X1（RNA聚合酶Ⅱ转录酶亚基15a亚型X1的介体）	玉米	gi 1011750777	Chr4	9.204×10^{-115}
M5/P12	T51^a	361	homeodomain–like transcription factor superfamily protein（同源结构域转录因子超家族蛋白）	玉米	gi 413936027	Chr5	4.426×10^{-147}
M2/P5	T117	631	jasmonate–zim–domain protein 1（转录抑制因子JAZ蛋白1）	拟南芥	gi 414885350	Chr7	2.145×10^{-172}

续表

引物组合	DEG	大小/bp	同源蛋白	物种	GenBank登录号	Chr^c	E值
M4/P2	T222	230	RING zinc finger domin superfamily protein（环锌指结构域超家族蛋白）	玉米	gi|413924986	Chr4	4.866×10^{-95}
M1/E6	T286	402	jumonji-like transcription factor family protein（jumonji 转录因子家族蛋白）	玉米	gi|414885308	Chr7	9.921×10^{-20}
M6/E8	T269	184	RING zinc finger domin superfamily protein（环锌指结构域超家族蛋白）	玉米	gi|413942672	Chr6	9.692×10^{-47}
M7/P6	T34	193	MOS1 modifier of snc1（免疫调控因子 MOS1）	拟南芥	gi|414868482	Chr1	3.253×10^{-41}
M9/P8	T170	368	protein SMG7（SMG7 蛋白）	玉米	gi|670378242	Chr7	5.476×10^{-116}
M7/P7	T38	419	NAC domain transcription factor superfamily protein（NAC 结构域转录因子超家族蛋白）	玉米	gi|414873581	Chr1	0
翻译和翻译后修饰(10)							
M7/P9	T1^a	465	ribosomal protein（核糖体蛋白）	玉米	gi|413945206	Chr6	2.974×10^{-60}
M7/P9	T2^a	464	ribosomal protein（核糖体蛋白）	玉米	gi|413945206	Chr6	2.974×10^{-60}
M4/P6	T290	134	ribosomal protein S7（核糖体蛋白 S7）	玉米	gi|111467245	ChrPt	1.030×10^{-29}

续表

引物组合	DEG	大小/bp	同源蛋白	物种	GenBank 登录号	Chr[c]	E 值
M7/P12	T11	161	chaperone protein dnaJ（伴侣蛋白 dnaJ）	玉米	gi|414870267	Chr1	2.176×10^{-12}
M9/P1	T147[b]	169	ion protease‐like protein 2（离子蛋白酶 2）	玉米	gi|414886455	Chr7	6.010×10^{-58}
M6/P3	T194[a]	197	methionine aminopeptidase（甲硫氨酸氨肽酶）	玉米	gi|413921048	Chr4	2.457×10^{-92}
M9/P2	T155	301	serine peptidase S28 family protein（丝氨酸肽酶 S28 家族蛋白）	玉米	gi|414870775	Chr1	1.697×10^{-145}
M7/P7	T39	281	Rhomboid‐like protein 9（Rhomboid 蛋白 9）	小米	gi|413925970	Chr4	4.866×10^{-71}
M8/E6	T259	307	ATP‐dependent Clp protease proteolytic subunit isoform 1（ATP 依赖的丝氨酸 Clp 酶蛋白水解亚基亚型 1）	玉米	gi|414866760	Chr1	4.448×10^{-22}
M2/P2	T85	481	nascent polypeptide‐associated complex alpha sub-unit‐like protein（新生多肽相关复合物 α 亚基蛋白）	玉米	gi|414866831	Chr1	9.150×10^{-31}
推测及未分类蛋白（6）							
M8/P9	T50	182	hypothetical protein ZEAMMB73_234053（推测蛋白 ZEAMMB73_234053）	玉米	gi|414586826	Chr2	8.781×10^{-32}

续表

引物组合	DEG	大小/bp	同源蛋白	物种	GenBank 登录号	Chr[c]	E 值
M8/E6	T260	212	hypothetical protein ZEAMMB73_898558（推测蛋白 ZEAMMB73_898558）	玉米	gi|413937220	Chr5	4.587×10^{-105}
M6/P2	T190	254	hypothetical protein ZEAMMB73_762270（推测蛋白 ZEAMMB73_762270）	玉米	gi|414872780	Chr1	9.095×10^{-93}
M8/E6	T256	439	putative DUF2451 domain containing family protein（推测含 DUF2451 结构域的家族蛋白）	玉米	gi|414867616	Chr1	0
M6/E7	T265	462	hypothetical protein ZEAMMB73_841898（推测蛋白 ZEAMMB73_841898）	玉米	gi|413918397	Chr10	4.939×10^{-108}
M3/E1	T292	87	hypothetical protein ZEAMMB73_434839（推测蛋白 ZEAMMB73_434839）	玉米	gi|413926462	Chr4	3.775×10^{-27}
无显著相似性蛋白（8）							
M7/P1	T20	277	no hit（无相似性序列）	—	—	—	—
M4/P2	T223	205	no hit	—	—	—	—
M4/P3	T227	212	no hit	—	—	—	—
M8/E6	T261	238	no hit	—	—	—	—
M10/E7	T291	216	no hit	—	—	—	—
M7/E2	T243	110	no hit	—	—	—	—
M3/E6	T293	126	no hit	—	—	—	—
M3/E8	T297	138	no hit	—	—	—	—

注：a. 冷胁迫后在玉米叶片和根部同时上调的 DEG；b. 冷胁迫后在玉米叶片和根部同时下调的 DEGs；c. DEG 在玉米基因组中的染色体定位。

2.2.6　差异表达基因的验证

为了进一步验证 cDNA-AFLP 获得的候选基因是否在冷胁迫下真实差异表达,随机挑选 16 个 DEG,同时利用 real-time PCR 对其进行相对定量分析,结果如图 2-9 和图 2-10 所示。两种方法的基因表达结果趋势一致,表明 cDNA-AFLP 结果是可靠的。

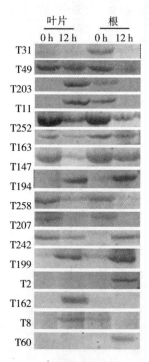

图 2-9　DEG 分析(cDNA-AFLP)

第 2 章　冷胁迫下玉米叶片和根部基因表达谱的研究

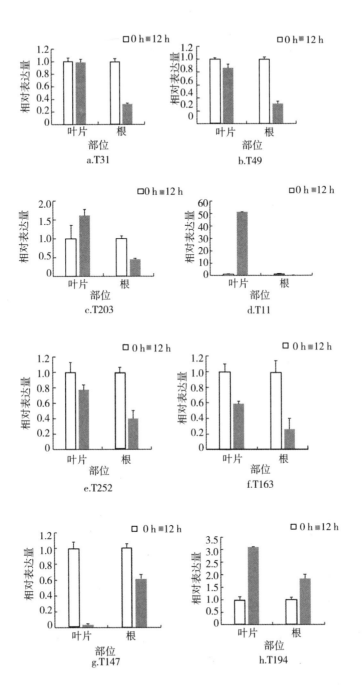

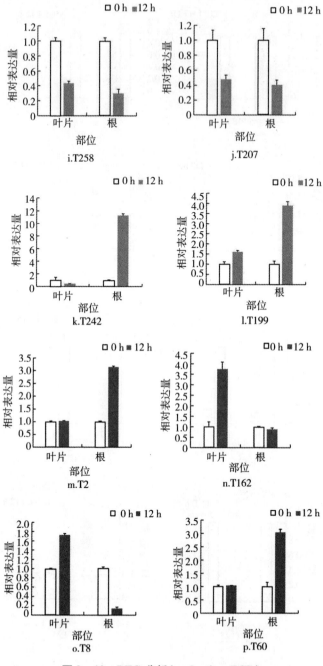

图 2-10 DEG 分析(real-time PCR)

2.3 讨论

cDNA-AFLP 是一种 mRNA 指纹图谱技术,其将 real-time PCR 和 AFLP 技术相结合,可对生物体转录组进行全面分析。同时,该技术保留了 AFLP 技术的稳定性高、多态性丰富、无须了解序列信息等优点,集中显示基因组表达序列的多态性差异,且具有重复性好、准确、可靠等特点,已成功用于植物遗传标记分析、基因表达特性研究和分离植物 DEG 等方面。有研究人员利用 cDNA-AFLP 技术成功分离了玉米抗冷基因 *ZmCLC-d*,过表达 *ZmCLC-d* 拟南芥能显著提高转基因植株的抗冷性;利用此技术成功克隆了玉米体细胞胚胎形成相关基因 *ZmSUF4* 和 *ZmDRP3A*;成功构建了拟南芥和马铃薯表达基因的转录连锁图。cDNA-AFLP 不仅可以检测出缺失或插入引起的差异,同时也可检测到点突变造成的差异。

本章利用前期筛选的抗冷自交系 W9816 为材料,在冷胁迫下,利用 cDNA-AFLP 技术对玉米苗期叶片和根部的 DEG 进行检测和分离。在玉米叶片和根部分别检测到 620 个和 531 个 DEG。冷胁迫下,大多数 DEG 在玉米叶片和根部中表达趋势不一致,仅仅 61 个上调的 DEG 和 32 个下调的 DEG 在两种组织中具有相同的表达趋势。这些结果表明了基因空间调控的重要性以及在不同组织中存在不同的冷响应调控机制。对 67 个 DEG 成功测序,一些重要的 DEG 根据其功能讨论如下。

2.3.1 信号转导

植物通过一系列膜结合受体感知冷信号。钙调结合蛋白能够感知钙离子信号,并且通过与效应蛋白的相互作用来激活细胞响应。本试验中钙调结合蛋白(T203)在叶片中上调表达,在根部下调表达。富含亮氨酸重复受体在植物系统中属于一类类受体蛋白激酶,其功能参与感知与传递环境信号进而调节逆境响应。一个编码 LRR 的 DEG(T162)在冷胁迫玉米叶片中上调表达。有研究报道了一个新的冷激活富含亮氨酸重复受体大豆 GsLRPK,其可作为抗冷相关的正调控因子。磷脂信号级联反应在植物响应各种不利生长条件中发挥着举足轻重的作用。T287 编码一个 Sec14 类 PIPT,冷胁迫下此基因在玉米叶片和根部均上调表达。Sec14 类 PIPT 家族参与各种生物学过程，如信号转导、磷脂代谢和逆境响应等。

2.3.2 转录调控

许多研究表明转录因子在调控逆境相关基因的转录中具有重要的作用。在本章中，T51、T222 以及 T286 属于同源结构域转录因子、环锌指结构域超家族转录因子以及 jumonji 类转录因子，其转录受冷胁迫上调表达。NMD 参与 mRNA 的监督机制，保护机体免遭非功能或者有害多肽积累的伤害。SMG7 结合磷酸化的 UPF1，UPF1 是激活 NMD 途径中最重要的因子。冷胁迫下玉米叶片 *SMG7* 基因转录的上调可能提高清除不正常 mRNA 的效率。

2.3.3 翻译和翻译后修饰

T194 编码甲硫氨酸氨肽酶，冷胁迫可上调玉米叶片和根部 T194 表达。拟南芥过表达 *HvMAP* 可能通过促进蛋白质成熟来提高植株抗冷性。Hsp70 在正常环境和胁迫环境下可以防止蛋白质凝聚和提高非天然蛋白质的重折叠，DnaJ 可以作为 Hsp70 蛋白的共伴侣。NAC 可以作为"分子伴侣"与未折叠多肽链相互作用。本研究中冷胁迫分别在玉米叶片和根部上调伴侣蛋白 DnaJ 和 NAC 的表达。

2.3.4 细胞代谢与组织

T8 和 T258 参与氨基酸合成代谢，其表达受冷胁迫影响。已有报道表明多胺在各种抗逆响应中起着关键作用。SAMD 是多胺生物合成的限速酶之一。低温诱导 SAMD 活性，冷处理 12 h 后，SAMD 活性的上升能够提高黄瓜的抗冷性。本研究发现 *SAMD* 的转录水平在冷胁迫玉米叶片中上调表达。谷氨酰胺-谷氨酸合成循环被认为是谷氨酸的最重要来源。谷氨酸是脯氨酸合成的一个前体。冷胁迫后 T258 的转录水平在玉米叶片和根部下调表达。mRNA 水平和相应蛋白质丰度之间的差异表明冷胁迫玉米叶片中存在转录后调控和翻译后修饰等过程。T246 和 T10 分别编码 NADH 脱氢酶 I 亚基 1 和过氧化物酶 51，其转录水平在冷胁迫玉米根部上调表达。非生物胁迫能够产生活性氧,其在高浓度下会引起不可逆转的代谢紊乱。过氧化物酶51的上调表达可能脱毒ROS和增强植株的抗冷性。

2.4 结论

(1) 采用cDNA-AFLP技术,利用174对引物组合,对玉米冷胁迫下不同组织的DEG进行了分析,发现4种不同的基因表达模式。

(2) 成功测序67个DEG,涉及植物生长发育的多个过程。

(3) real-time PCR验证了cDNA-AFLP结果的可靠性。

(4) 筛选出了Sec14类PIPT等冷响应相关的新基因。

第 3 章

玉米 ZmSEC14p 基因的克隆及序列分析

甘油酯是植物细胞膜的主要结构组分,在信号转导和膜运输中发挥重要的调控作用。其主要的合成位点是内质网,但是在植物中,叶绿体中也合成脂单体、脂肪酸,后转移到细胞质中参与不同脂类(如 PtdCho、磷脂酰乙醇胺、PtdIns、磷脂酰丝氨酸、PA、磷脂酰甘油和半乳糖)的生物合成。PtdIns 的肌醇环 3'、4' 和 5' 的位置能够被不同激酶/磷酸酶磷酸化,进而形成一系列 PtdIns 衍生物,包括磷脂酰肌醇-3-磷酸、磷脂酰肌醇-4-磷酸、磷脂酰肌醇-5-磷酸和 PIP2。

内质网和叶绿体产生的脂类需要通过有效的运输系统转移到细胞内的靶位点。真核生物细胞内脂转运存在两个主要的方式:一种是囊泡转运,主要是在叶绿体膜和内质网膜之间转运脂肪酸;另一种是通过脂转移蛋白进行调节。脂转移蛋白具有选择性的疏水口袋,可以允许一个或者多个脂与其结合。根据脂转移蛋白结合特异性可将其分为三类:甘油磷脂、鞘脂和固醇转移蛋白。PIPT 属于甘油磷脂转移蛋白,其内部疏水区能够结合 PIP。所有 PIPT 的原型均来源于酵母 Sec14p 蛋白,作为高尔基体装置的外周蛋白,在反式高尔基体亚室之间进行膜转运。越来越多的证据表明 PIPT 可调节 PIP 的转运以及膜内 PIP 存在的丰度。

最近的许多研究表明,PIPT 参与不同植物的非生物胁迫。棕藻酮胁迫短期驯化过程中,转录组与代谢组分析表明 PIPT Sec14 下调表达。高粱不同基因型转录谱研究表明抗冷高粱基因型中具有更高的 Sec14 转录本,这一变化可能增加膜的稳定性以及对氮的抗性。J. Lee 等报道冷胁迫下,抗冷卷心菜中 Sec14 相关基因具有更高的表达水平。Sec14 类蛋白质参与基本生物学过程,例如磷脂代谢、膜转运、极性膜生长、信号转导和逆境响应。尽管 PIPT 存在于所有真核生物中,但在玉米中关于 Sec14 类相关蛋白质的功能及生理学过程的研究还未见报道。本章中利用已获得的 DEG 信息,通过 RACE 成功克隆了 *ZmSEC14p* 基因全长,并对基因进行了生物信息学分析。

3.1 试验材料与方法

3.1.1 菌种与载体

菌种为大肠杆菌Top10,载体为pMD^{TM}18-T克隆载体。

3.1.2 主要试剂

RNA提取试剂(RNAiso Plus)、Oligo(dT)$_{18}$引物(50 μmol/L)、M-MLV逆转录酶(RNase H$^-$,200 U/μL)、dNTP(2.5mmol/L)、核糖核酸酶抑制剂(40 U/μL)、DL-2000 marker、RNase A、RNase H、末端脱氧核苷酸转移酶(TdT)、L*Taq*、PrimerSTAR HS DNA Polymerase、各种限制性内切酶、琼脂糖回收试剂盒、超薄DNA产物纯化试剂盒、氯化钙、Amp、琼脂糖、RNaseA、三羟甲基氨基甲烷、乙二胺四乙酸二钠、三氯甲烷、异丙醇、氢氧化钠、乙醇、SDS、蛋白胨、酵母提取物以及琼脂粉。

3.1.3 试剂及培养基的配制

试验所需试剂及培养基的配制同2.1.3。

3.1.4 主要仪器

试验所需主要仪器同2.1.4。

3.1.5 植物材料

以实验室筛选的玉米抗冷自交系W9816为试验材料,种植方法同2.1.5。

3.1.6 冷胁迫处理

冷胁迫处理同2.1.6。

3.1.7 RNA 提取和 cDNA 第一链的合成

RNA 的提取同 2.1.7，cDNA 第一链的合成同 2.1.8 和 2.1.9.1。

3.1.8 *ZmSEC14p* 3'端序列的克隆

采用 3'RACE 方法进行 *ZmSEC14p* 3'端的克隆，技术路线如图 3-1 所示。

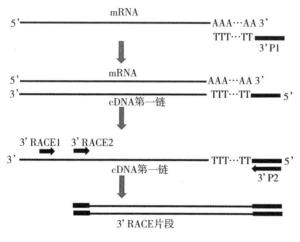

图 3-1　3'RACE 技术流程图

3.1.9 *ZmSEC14p* 3'端引物设计与合成

利用 cDNA-AFLP 分析获得的 T287 序列，设计两个特异性上游引物 3'RACE1、3'RACE2；下游引物根据 mRNA 3'端具有 Poly A 尾结构，设计了反转录锚定引物 3'P1 和 3'P2，序列见表 3-1。

表 3-1 引物

引物名称	引物序列 (5'—3')
3'RACE1	TGAAGAAAGTCTCAAATGGAGGGCA
3'RACE2	TGACGGTCAGATTCGGTTTCTTGTG
3'P1	TGCGTGGAGGACATTGTGGTAGTG(dT)$_{18}$
3'P2	TGCGTGGAGGACATTGTGGTAGTG

3.1.10 *ZmSEC14p* 3'端的扩增

利用反转录引物 3'P1 合成的 cDNA 为模板,分别用 3'RACE1、3'RACE2 与 3'P2 进行巢式 PCR,反应体系见表 3-2。

表 3-2 巢式 PCR 反应体系 (25 μL)

成分	用量/μL
cDNA	2
10×LA PCR buffer	2.5
dNTP(2.5 mmol/L)	2
3'RACE2(10 μmol/L)	0.5
3'P1(P2)(10 μmol/L)	0.5
L*Taq*(5 U/μL)	0.2
ddH$_2$O	17.3

反应条件如下:

94 ℃ 3 min

94 ℃ 30 s
55 ℃ 30 s } 20 个循环(第一轮反应)
72 ℃ 2 min   25 个循环(第二轮反应)

72 ℃ 10 min

4 ℃ 终延伸

3.1.11 ZmSEC14p 3'端 PCR 产物的回收、连接及测序

目的片段的琼脂糖凝胶回收同 2.1.12,目的片段和克隆载体的连接同 2.1.13.1,重组质粒的转化和 PCR 鉴定分别同 2.1.13.3 和 2.1.13.4。将 PCR 鉴定结果为阳性的菌液测序,至少检测 3 个阳性克隆以减少 PCR 造成的误差。

3.1.12 ZmSEC14p 5'端的克隆

利用 5'RACE 方法进行 ZmSEC14p 5'端序列的扩增,技术路线如图 3-2 所示。

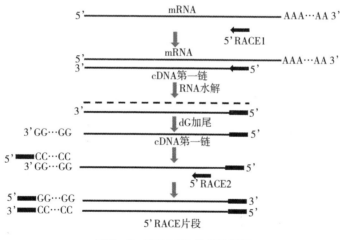

图 3-2 5'RACE 技术流程图

3.1.13 ZmSEC14p 5'端引物的设计与合成

根据 3'RACE 获得的基因序列,设计特异性下游引物 5'RACE1、5'RACE2。由于 cDNA 3'端进行了 Poly G 加尾反应,所以根据 Poly G 序列设计 Poly C 序列 5'上游引物,序列见表 3-3。

表 3-3　引物

引物名称	引物序列（5'—3'）
5'RACE1	CAATGGCCAGCCTCTCAGGGTAATG
5'RACE2	CATACACAAGAAACCGAATCTGACC
5'P	TTAAATTAATCCCCCCCCCCCCCCC

3.1.14　cDNA 第一链的合成

RNase H 被用于分解 RNA–cDNA 杂交体中的 RNA，采用 RNaseA 降解单链 RNA，反应体系见表 3-4。

表 3-4　反应体系（50 μL）

成分	用量/μL
cDNA	38
5×RNA 杂交 buffer	10
RNase H(60 U/μL)	1
核糖核酸酶 A(10 mg/mL)	1

30 ℃反应 1 h，37 ℃反应 1 h。

3.1.15　cDNA 第一链的纯化

采用超薄 DNA 产物纯化试剂盒进行纯化。

(1) 向吸附柱 CB1（吸附柱放入收集管中）中加入 500 μL 的平衡液 BL（由试剂盒提供），12 000 r/min（约 13 400 ×g）离心 1 min，倒掉收集管中的废液，将吸附柱重新放回收集管中。

(2) 向 50 μL 含有降解 RNA 的 cDNA 混合物中加入 250 μL 的结合液 PB（由试剂盒提供），充分混合。

(3) 将上一步所得溶液加入吸附柱 CB1（吸附柱放入收集管中）中，室温放置 2 min，12 000 r/mim（约 13 400 ×g）离心 30 s，倒掉收集管中的废液，将吸附柱 CB1 放入收集管中。

（4）向吸附柱 CB1 中加入 600 μL 漂洗液 PW（由试剂盒提供），12 000 r/min（约 13 400×g）离心 30 s，倒掉收集管中的废液，将吸附柱 CB1 放入收集管中。

（5）重复操作步骤(4)。

（6）12 000 r/min（约 13 400×g）离心 2 min，尽量除去漂洗液。将吸附柱于室温放置数分钟，彻底晾干，以防止残留的漂洗液影响下一步试验。

（7）取出吸附柱 CB1，放入一个干净的离心管中，向吸附膜中间位置悬空滴加 30 μL 洗脱缓冲液 EB（由试剂盒提供），室温放置 2 min。12 000 r/min（约 13 400×g）离心 2 min，收集 DNA 溶液。重复此步骤可提高回收效率。

3.1.16 纯化 cDNA 第一链的加尾反应

利用 TdT 对 cDNA 第一链的 3'末端进行 Poly G 加尾反应，反应体系见表 3-5。

表 3-5 加尾反应体系（50 μL）

成分	用量/μL
cDNA	20
0.1% BSA	5
5×TdT buffer	10
dGTP(100 mmol/L)	0.5
TdT(7~15 U/μL)	1
ddH$_2$O	13.5

37 ℃反应 30 min。

3.1.17 加尾 cDNA 的纯化

加尾 cDNA 的纯化同 3.1.15。

3.1.18 *ZmSEC14p* 5'端的克隆

利用反转录合成的 cDNA 为模板,5'RACE1、5'RACE2 分别与 5'P 进行巢式 PCR,反应体系见表 3-6。

表 3-6 巢式 PCR 反应体系(25 μL)

成分	用量/μL
cDNA	2
10×PCR buffer	2.5
dNTP(2.5 mmol/L)	2
5'RACE1(RACE2)(10 μmol/L)	0.5
5'P(10 μmol/L)	0.5
L*Taq*(5 U/μL)	0.2
ddH$_2$O	17.3

反应条件为:
94 ℃ 3 min
94 ℃ 30 s ⎫
55 ℃ 30 s ⎬ → { 20 个循环(第一轮反应)
72 ℃ 2 min ⎭   { 25 个循环(第二轮反应)
72 ℃ 10 min
4 ℃ 终延伸

3.1.19 *ZmSEC14p* ORF 的克隆

利用 DNAman 软件对 3'RACE 和 5'RACE 克隆序列进行拼接,通过 NCBI ORF Finder 寻找 *ZmSEC14p* 基因完整的 ORF,进而根据 ORF 序列设计特异性引物,序列见表 3-7。

表 3-7 ORF 克隆的特异性引物

基因	引物序列(5'—3')		T_m/℃
ZmSEC14p	上游	ATGTTCAGGAGAAAGCATGCTTCTC	56
	下游	TCATTAACTGGCTTTACAGCAATC	

以合成的 cDNA 第一链为模板，利用高保真酶 PrimerSTAR HS DNA Polymerase 扩增 ZmSEC14p 基因的 ORF，PCR 体系见表 3-8。

表 3-8 反应体系(25 μL)

成分	用量/μL
cDNA	2
5×PrimerSTAR buffer(Mg^{2+} Plus)	2
dNTP(2.5 mmol/L)	2
PCR 引物序列(F)(10 μmol/L)	0.5
PCR 引物序列(R)(10 μmol/L)	0.5
PrimerSTAR HS DNA Polymerase(2.5 U/μL)	0.5
ddH$_2$O	17.5

反应条件：

98 ℃ 10 s
55 ℃ 15 s } 30 个循环
72 ℃ 1 min

3.1.20 回收产物的平末端加 A

加 A 反应体系见表 3-9。

表 3-9 加 A 反应体系

成分	用量
末端平滑 DNA 片段	0.5~5 μg
dNTP 混合溶液	4 μL
A - 加尾酶	0.5 μL
ddH$_2$O	加至体系共 50 μL

72 ℃反应 20 min,冰上静置 1~2 min。

3.1.21　琼脂糖凝胶回收、连接及测序

目的片段的琼脂糖凝胶回收同 2.1.12,目的片段和克隆载体的连接同 2.1.13.1,重组质粒的转化和 PCR 鉴定分别同 2.1.13.3 和 2.1.13.4。将 PCR 鉴定结果为阳性的菌液测序,至少检测 3 个阳性克隆以减少 PCR 造成的误差。

3.1.22　生物信息学分析

使用 NCBI Conserved Domains(http://www.ncbi.nlm.nih.gov/Structure/cdd/wrpsb.cgi)预测基因保守区;利用 ExPASy Proteomics Server(http://www.expasy.ch/too ls/protparam.html)预测蛋白质等电点、分子质量等;蛋白质多重序列比对、进化树分析分别利用 DNAman 和 MEGA7 软件;ZmSEC14p 蛋白三维结构利用 I - TASSER 在线资源(http://zhanglab.ccmb.med.umich.edu/I - TASSER)进行预测。

3.2　结果与分析

3.2.1　玉米 *ZmSEC14p* 3'端序列的克隆

本章利用锚定引物 3'P1 进行反转录,3'RACE1 和 3'RACE2 分别与 3'P2 进行巢式 PCR,扩增得到 632 bp 的目的条带,结果如图 3-3 所示。

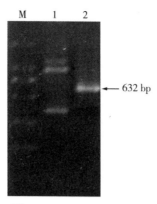

图 3-3 3'RACE 扩增结果
1.3'RACE 第一轮扩增;2.3'RACE 第二轮扩增;
M. DL-2000 marker

3.2.2 玉米 *ZmSEC14p* 5'端序列的克隆

利用 Oligo(dT)$_{18}$ 引物进行反转录合成 cDNA,经过 RNA 分解,cDNA 第一链的纯化、加尾等步骤,通过 5'RACE1、5'RACE2 和 5'P 分别进行巢式 PCR,最终扩增得到 741 bp 的目的条带,结果如图 3-4 所示。

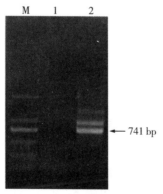

图 3-4 5'RACE 扩增结果
1.5'RACE 第一轮扩增;2.5'RACE 第二轮扩增;
M. DL-2000 marker

3.2.3 玉米 *ZmSEC*14*p* ORF 克隆

对 *ZmSEC*14*p* 3'端和 5'端测序结果进行分析,通过 DNAman 软件进行序列拼接,成功拼接到 1 346 bp 的全长 cDNA。利用 NCBI 的 ORF Finder 进行分析,结果表明 *ZmSEC*14*p* 包含 888 bp 的完整 ORF,推测编码 295 个氨基酸,同时含有 251 bp 的 5'UTR 和 207 bp 的 3'UTR 区。染色体定位表明 *ZmSEC*14*p* 定位于 1 号染色体,横跨 3 420 bp,包括 5 个外显子。如图 3-5、图 3-6 所示。

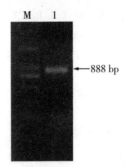

图 3-5 *ZmSEC*14*p* ORF 扩增结果

M. DL-2000 marker;1. *ZmSEC*14*p*

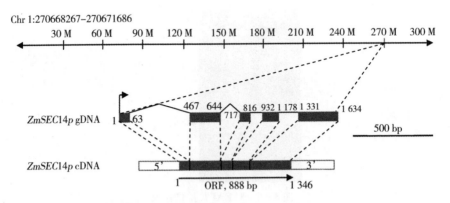

图 3-6 *ZmSEC*14*p* 在染色体中的定位

3.2.4 ZmSEC14p 蛋白的结构特性

ZmSEC14p 蛋白推测的分子质量为 34.12 ku,等电点为 7.09。295 个氨基酸中,50 个是碱性氨基酸(+ ,R、K、H),42 个是酸性氨基酸(- ,D、E);118 个是疏水氨基酸(A、V、L、I、P、F、W、M),85 个是极性氨基酸(G、S、T、C、Y、N、Q);42 个是极性带负电氨基酸(D、E),42 个是极性带正电氨基酸(R、K)。利用 ProtParam 工具(http://ca.expasy.org/tools/protparam.html)计算 ZmSEC14p 的不稳定指数是 42.44,表明该蛋白质为不稳定蛋白。利用 NCBI Conserved Domains(http://www.ncbi.nlm.nih.gov/Structure/cdd/wrpsb.cgi)对基因的保守区进行预测,结果如图 3 - 7 所示。分析表明 ZmSEC14p 在 86 ~ 240 个氨基酸处含有保守的 SEC14 结构域,氨基酸残基潜在形成了 ZmSEC14p 的磷脂结合口袋(E113、S115、Q122、I122、W145、I147、F149、W152、H156、P159、T162、A163、C166、T167、A181、F190、F193、Y194、V197、L209),N 末端包含一个保守的 CRAL - TRIO - N(19 ~ 66)结构域。这表明 ZmSEC14p 属于典型的 Sec14 类蛋白质家族成员。二级结构预测表明该蛋白质包含 10 个 α 螺旋、5 个 β 折叠和 3 个 3_{10} 螺旋(图 3 - 8)。利用 I - TASSER 在线资源预测 ZmSEC14p 三维结构,I - TASSER 通常为每个任务提供多达 5 个全长原子模型,对于每个建模过程,显示最高估计精确度的模型将被考虑。I - TASSER 结构预测表明 ZmSEC14p 蛋白包含 7 个 α 螺旋、5 个 β 折叠以及 2 个 3_{10} 螺旋,可形成磷脂结合口袋(图 3 - 9)。

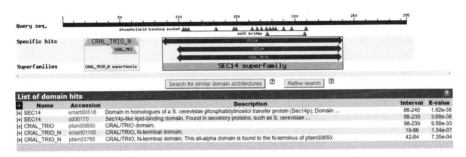

图 3 - 7 ZmSEC14p 保守区的预测

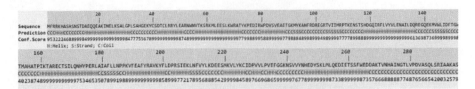

图 3-8　基于序列的二级结构预测

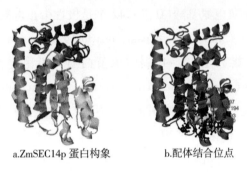

a. ZmSEC14p 蛋白构象　　　b. 配体结合位点

图 3-9　ZmSEC14p 蛋白的三维模型

3.2.5　多重序列比对和系统进化树分析

利用 DNAman 软件对多种植物(高粱、谷子、二穗短柄草、小麦、水稻和大麦)的同源 PIPTs 的氨基酸序列进行比对,同源性分别为 90.85%、85.42%、73.29%、73.29%、72.88% 和 71.58%(图 3-10)。参考相关研究可将 Sec14 类蛋白质分为 4 类:Sec14 同源组(SFH)、Patellin 组(PATL)、大豆 Sec14 同源组(SSH)以及未识别的同源组(UCSH)。试验发现 ZmSEC14p 属于 UCSH,仅仅包含一个 SEC14 结构域,进化树分析表明 ZmSEC14p 与高粱中的同源基因亲缘关系最近(图 3-11、图 3-12)。

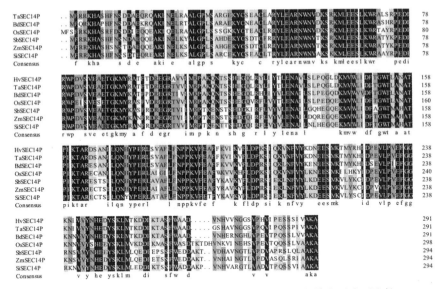

图 3-10　ZmSEC14p 与其他植物 Sec14p 同源蛋白的多重序列比较

注：基因来源和 GenBank 登录号为玉米（ZmSEC14p，KT932998），高粱（SbSEC14p，Sb01g009685.2），谷子（SiSEC14p，Si036849m.g），水稻（OsSEC14p，NP_001051120），二穗短柄草（BdSEC14p，BRADI1G10270），大麦（HvSEC14p，MLOC_64122），小麦（TaSEC14p，Traes_4DS_DB674AE37）。

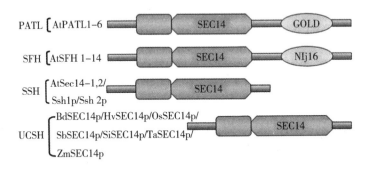

图 3-11　植物 Sec14 类蛋白质结构域示意图

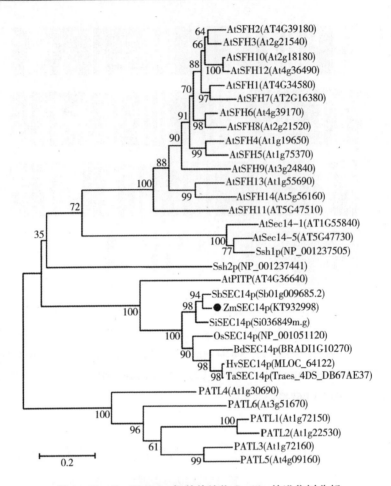

图 3-12 ZmSEC14p 与其他植物 Sec14p 的进化树分析

注：基因来源和 GenBank 登录号为玉米（ZmSEC14p, KT932998），拟南芥（AtPITP, AT4G36640；AtSec14-1，AT1G55840；AtSec14-5，AT5G47730；AtSFH1，AT4G34580；AtSFH2，AT4G39180；AtSFH3，At2g21540；AtSFH4，At1g19650；AtSFH5，At1g75370；AtSFH6，At4g39170；AtSFH7，AT2G16380；AtSFH8，At2g21520；AtSFH9，At3g24840；AtSFH10，At2g18180；AtSFH11，AT5G47510；AtSFH12，At4g36490；AtSFH13，At1g55690；AtSFH14，At5g56160；PATL1，At1g72150；PATL2，At1g22530；PATL3，At1g72160；PATL4，At1g30690；PATL5，At4g09160；PATL6，At3g51670），高粱（SbSEC14p, Sb01g009685.2），大豆（Ssh1p，NP_001237505；Ssh2p，NP_001237441），谷子（SiSEC14p, Si036849m.g），水稻（OsSEC14p，NP_001051120），二穗短柄草（BdSEC14p，BRADI1G10270），大麦（HvSEC14p, MLOC_64122），小麦（TaSEC14p，Traes_4DS_DB674AE37）。标尺为核苷酸替代率。

3.3 讨论

本章成功克隆了一个玉米 Sec14 PIPT 基因 *ZmSEC14p*。通过与玉米基因组的比较发现，*ZmSEC14p* 基因定位于 1 号染色体。玉米基因组中存在两个可变剪切体（GRMZM2G704053_T01、GRMZM2G704053_T02），其差别在于 5'UTR 和 1 号内含子的保留。通过 RACE 和 real-time PCR 的方法，在冷处理样本中仅发现了一个转录本，表明低温影响 *ZmSEC14p* 基因转录本的剪切方式。

PIPT 根据一级序列和结构折叠可分为两个不同的分支：Sec14 类 PIPTs 以及 START 类 PIPT。真菌和植物 PIPT 属于 Sec14 类 PIPT，后生动物属于 START 类 PIPT。人类 7 个 START 类 PIPT 根据进化分析可分为两类：Class Ⅰ，包括 PIPTα 和 PIPTβ，其仅仅有一个 PIPT 结构域能够结合 PtdIns 或 PtdCho；Class Ⅱ 包括两个亚家族，即 Rdgα 和 Rdgβ，Class Ⅱ 成员除了能够结合 PtdIns 和 PtdCho，还能结合 PA。Sec14 类 PIPT 蛋白构成一个复杂的家族，其中一些成员包括额外的结构域，能够锚定供体和受体膜。酵母表达 6 个 Sec14 类蛋白质，包括原始 SEC14p 蛋白和 5 个其他的 SFH 蛋白。在植物当中，Sec14 类 PIPT 蛋白可被分为 4 组，所有成员在其 N 端均包含一个保守的 SEC14 结构域。拟南芥中，至少存在 31 个 Sec14 类基因，12 个基因编码 Sec14-nodulin PIPT，目前该组的极少数成员已被鉴定。拟南芥中 12 个成员属于 Sec14-GOLD 结构域 PIPT，主要代表是其中的 6 个 patellins 蛋白（PATL1-6）。植物 Sec14 类蛋白质的第三个家族仅包含一个 PIPT 结构域，到目前为止，这类基因的表达和功能分析的相关研究极少。烟草 SEC14 磷脂转运蛋白可能通过依赖茉莉酸防御信号途径来调控植物对假单胞菌的抗性。本章中系统进化树分析表明 ZmSEC14p 属于 UCSH 的一个成员，仅仅包含一个 SEC14 结构域。I-TASSER 结构预测表明 ZmSEC14p 蛋白包括 10 个 α 螺旋、5 个 β 折叠和 3 个 3_{10} 螺旋，其中 7 个 α 螺旋、5 个 β 折叠以及 2 个 3_{10} 螺旋可形成磷脂结合口袋。

3.4 结论

(1)利用 RACE-PCR 方法成功克隆了 *ZmSEC14p* 基因,全长包含 888 bp 的完整 ORF,251 bp 的 5' UTR 以及 207 bp 的 3' UTR;染色体定位表明 *ZmSEC14p* 定位于玉米基因组 1 号染色体上,横跨 3 420 bp,包括 5 个外显子。

(2)多重序列比对表明,玉米 ZmSEC14p 与高粱 SbSEC14p 氨基酸同源性最高;进化树分析表明,ZmSEC14p 属于 UCSH 的一个成员,仅仅包含一个 SEC14 结构域。

(3)I-TASSER 结构预测表明 ZmSEC14p 蛋白包括 10 个 α 螺旋、5 个 β 折叠和 3 个 3_{10} 螺旋,其中 7 个 α 螺旋、5 个 β 折叠以及 2 个 3_{10} 螺旋可形成磷脂结合口袋。

第 4 章

玉米 *ZmSEC14p* 基因的功能分析

到目前为止,植物中对仅含单一的 Sec14p 结构域的 PIPT 的研究极少。有研究人员报道大豆 *Sshp*1 和 *Sshp*2 的转录水平受发育调控,且 Sshp1 和 Sshp2 蛋白在体外与 PIP 的结合特异性稍有区别。烟草叶片接种青枯雷尔氏菌后,烟草 *NbSec*14 基因转录水平上调表达,同时 NbSec14 能够恢复酵母 *Sec*14*p* 温度敏感型突变体的生长缺陷以及受损的分泌转化酶的活性。A. Kielbowicz-Matuk 等在抗旱大麦基因型中识别了一个 *HvSec*14*p* 基因,在种子形成与萌发的特定发育阶段以及不同渗透胁迫下,*HvSec*14 基因在转录水平与蛋白质水平均上调表达。有研究人员在普通小麦中克隆了 *TaSEC*14*p*-5 基因,该基因在小麦孕穗期的不同组织中均有表达,同时该基因受盐、ABA、干旱以及低温逆境胁迫诱导表达。本章将获得的 *ZmSEC*14*p* 基因全长在拟南芥中进行过表达,并对 *ZmSEC*14*p* 基因增强转基因植株的抗冻性机制进行探讨。

4.1 试验材料

4.1.1 菌种和载体

载体为植物过表达载体 pCHF3300、亚细胞定位载体 pCG3300-eGFP (图 4-1),菌种为大肠杆菌 Top10、根癌农杆菌 EHA105。

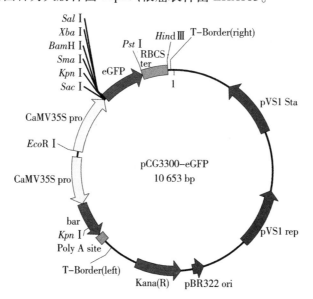

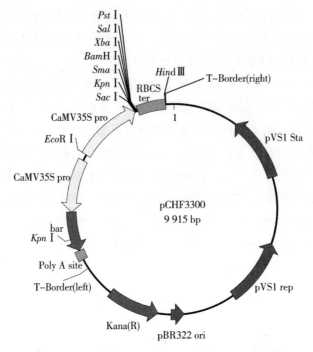

图 4-1　pCG3300-eGFP、pCHF3300 载体图谱

4.1.2　主要试剂

RNA 提取试剂（RNAiso Plus）、Oligo（dT）$_{18}$ 引物（50 μmol/L）、M-MLV 逆转录酶（RNase H$^-$，200 U/μL）、dNTP（2.5 mmol/L）、核糖核酸酶抑制剂（40 U/μL）、rTaq DNA 聚合物（5 U/μL）、DL-2000 marker、DL-15000 marker、PrimerSTAR HS DNA Polymerase、PrimeScript^{TM} RT Master Mix、SYBR Premix Ex Taq Ⅱ（Tli RNase H Plus）、各种限制性内切酶、琼脂糖回收试剂盒、氯化钙、Kana、Amp、琼脂糖、RNaseA、三羟甲基氨基甲烷、乙二胺四乙酸二钠、三氯甲烷、异丙醇、异戊醇、Tris-饱和酚、氢氧化钠、乙醇、SDS、蛋白胨、酵母提取物、琼脂粉、牛肉膏、蔗糖、硝酸钾、硝酸铵、磷酸二氢钾、硫酸镁、碘化钾、硼酸、硫酸锰、硫酸锌、钼酸钠、硫酸铜、氯化钴、硫酸亚铁、肌醇、甘氨酸、盐酸硫胺素、盐酸吡哆醇、烟碱。

4.1.3 主要仪器

试验所需主要仪器同 2.1.4。

4.1.4 试剂及培养基的配制

4.1.4.1 YEP 培养基

蛋白胨(10 g/L),酵母提取物(5 g/L),氯化钠(5 g/L),调 pH 值至 7.2。

4.1.4.2 2×CTAB 提取液

CTAB(0.02 g/mL),氯化钠(1.4 mol/L),EDTA(20 mmol/L),Tris-HCl(100 mmol/L,pH 值为 8.0),使用前加入 2% β-巯基乙醇。

4.1.4.3 MS 培养基

MS 培养基配方见表 4-1,加 1 mol/L 的氢氧化钠,调 pH 值为 5.8~6.0。

表 4-1 MS 培养基配方

类别	成分	使用浓度/(mg·L^{-1})
大量元素	硝酸钾	1 900
	硝酸铵	1 650
	磷酸二氢钾	170
	硫酸镁	370
	氯化钙	440

续表

类别	成分	使用浓度
微量元素	碘化钾	0.83
	硼酸	6.2
	硫酸锰	22.3
	硫酸锌	8.6
	钼酸钠	0.25
	硫酸铜	0.025
	氯化钴	0.025
铁盐	乙二胺四乙酸二钠	37.3
	硫酸亚铁	27.8
有机成分	肌醇	100
	甘氨酸	2
	盐酸硫胺素	0.1
	盐酸吡哆醇	0.5
	烟碱	0.5
	蔗糖	$30\,000 \times 10^4$
	琼脂	$7\,000 \times 10^3$

4.1.4.4 拟南芥营养液

母液Ⅰ(100×):取 25.25 g 硝酸钾、24.65 g 硫酸镁溶解,定容至 500 mL。

母液Ⅱ(100×):取 23.6 g 硝酸钙溶解,定容至 500 mL。

母液Ⅲ(100×):取 0.69 g 硫酸亚铁、0.93 g 乙二胺四乙酸二钠分别溶解,混合后定容至 500 mL。

母液Ⅳ(1 000×):取 4 340 mg 硼酸、2 772 mg 氯化锰、125 mg 硫酸铜、287.5 mg 硫酸锌、72.6 mg 钼酸钠、58.5 mg 氯化钠、2.38 mg 氯化钴溶解,定容至 1 L。

母液Ⅴ(100×):取 17 g 磷酸二氢钾溶解,定容至 500 mL。

取母液Ⅰ、母液Ⅱ、母液Ⅲ、母液Ⅴ各 10 mL,母液Ⅳ 1 mL,加纯化水定容到 1 L,调 pH 值至 5.7。

4.2 试验方法

4.2.1 两周龄玉米幼苗的处理

冷处理:将待处理的玉米幼苗放入 4 ℃ 人工气候箱中进行不同时间的冷处理,分别取幼苗叶片和根部,5 株混样后,液氮速冻,保存于 -80 ℃ 冰箱。

盐处理:用 250 mmol/L 的氯化钠完全浸没玉米幼苗,进行不同时间的盐处理,为了防止植物由于无氧呼吸受到损伤,处理期间要对玉米幼苗根部进行通气。分别取玉米幼苗叶片和根部,5 株混样后,液氮速冻,保存于 -80 ℃ 冰箱。

ABA 处理:用 100 μmol/L 的 ABA 喷洒玉米幼苗,进行不同时间的 ABA 处理。分别取玉米幼苗的叶片和根部,5 株混样后,液氮速冻,保存于 -80 ℃ 冰箱。

分别取玉米 W9816 孕穗期的不同组织(根部、茎、叶片、未成熟的雄穗、花药、穗轴、穗上叶、花丝和气生根),提取不同组织的 RNA,保存于 -80 ℃ 冰箱,用于组织特异性表达分析。

4.2.2 *ZmSEC14p* 表达载体的构建

4.2.2.1 *ZmSEC14p* 过表达载体的构建

植物过表达载体 pCHF3300 - *ZmSEC14p* 的构建策略如图 4 - 2 所示。

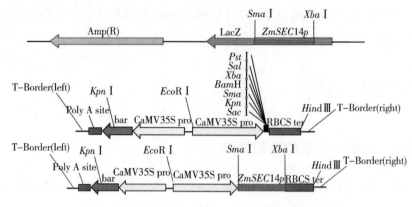

图4-2 pCHF3300-*ZmSEC*14*p* 过表达载体的构建策略

（1）用高保真酶扩增 *ZmSEC*14*p* 基因的 CDS 序列。根据载体中提供的限制性内切酶位点，在引物设计时加入合适的酶切位点（表4-2）。

表4-2 酶切位点与基因克隆引物

基因		引物序列(5'—3')	备注
*ZmSEC*14*p*	上游	TCCCCCGGGATGTTCAGGAGAAAGCATGCTTCTC	*Sma* I 酶切位点
	下游	CTAGTCTAGATCATTAACTGGCTTTACAGCAATC	*Xba* I 酶切位点

注：斜体部分对应酶切位点。

（2）扩增出的目的基因，经过回收、加 A 反应、T 载体连接、转化、酶切鉴定以及测序等一系列步骤，确定基因具有完整的 ORF。

（3）利用设计的内切酶切割 pMD™18-T 载体，回收基因小片段。

（4）以相同的内切酶切割 pCHF3300 过表达载体，回收载体大片段。

（5）将回收的大、小片段进行连接。

（6）转化大肠杆菌 Top10 感受态细胞，提取质粒进行 PCR 和酶切鉴定。

4.2.2.2 目的基因的克隆

使用克隆基因 CDS 的引物，在其两端分别加上 *Sma* I 和 *Xba* I 酶切位点，克隆带有酶切位点的 ORF，扩增体系与条件同 3.1.19，引物序列见表4-2。

4.2.2.3 目的基因的琼脂糖回收

目的基因的琼脂糖回收同 2.1.12。

4.2.2.4 目的片段与克隆载体连接

连接体系和条件同 2.1.13.1。

4.2.2.5 重组质粒的转化

重组质粒的转化同 2.1.13.3。

4.2.2.6 重组质粒的 PCR 鉴定

重组质粒 PCR 鉴定体系同 2.1.13.4。

4.2.2.7 质粒的提取

采用质粒小提试剂盒,方法如下。

(1)柱平衡步骤:向吸附柱 CP3(吸附柱放入收集管中)中加入 500 μL 的平衡液 BL(由试剂盒提供),12 000 r/min(约 13 400 ×g)离心 1 min,倒掉收集管中的废液,将吸附柱重新放回收集管中。

(2)取 1~5 mL 过夜培养的菌液,加入离心管中,12 000 r/min(约 13 400 ×g)离心 1 min,尽量吸除上清液(菌液较多时可以通过多次离心将菌体沉淀收集到一个离心管中)。

(3)向留有菌体沉淀的离心管中加入 250 μL 溶液 P1(由试剂盒提供,请先检查是否已加入 RNaseA),使用移液器或旋涡振荡器彻底悬浮细菌沉淀。

(4)向离心管中加入 250 μL 溶液 P2(由试剂盒提供),温和地上下翻转 6~8 次,使菌体充分裂解。

(5)向离心管中加入 350 μL 溶液 P3(由试剂盒提供),立即温和地上下翻转 6~8 次,充分混匀,此时将出现白色絮状沉淀。12 000 r/min(约 13 400 ×g)离心 10 min。

(6)将上一步收集的上清液用移液器转移到吸附柱 CP3(吸附柱放入收集管中)中,注意尽量不要吸出沉淀。12 000 r/min(约 13 400 ×g)离心

30~60 s,倒掉收集管中的废液,将吸附柱 CP3 放入收集管中。

(7)向吸附柱 CP3 中加入 600 μL 漂洗液 PW(由试剂盒提供,请先检查是否已加入无水乙醇),12 000 r/min(约 13 400×g)离心 30~60 s,倒掉收集管中的废液,将吸附柱 CP3 放入收集管中。重复操作此步骤。

(8)将吸附柱 CP3 放入收集管中,12 000 r/min(约 13 400×g)离心 2 min,目的是将吸附柱中残余的漂洗液去除。

(9)将吸附柱 CP3 置于一个干净的离心管中,向吸附膜的中间部位滴 50~100 μL 洗脱缓冲液 EB,室温放置 2 min,12 000 r/min(约 13 400×g)离心 2 min,将质粒溶液收集到离心管中。

4.2.2.8 重组质粒的酶切鉴定

重组质粒的酶切体系见表 4-3。

表 4-3 酶切体系

成分	用量
质粒 DNA	1 μg
10×M buffer	2 μL
Sma I (10 U/μL)[Xba I (15 U/μL)]	1 μL
1% BSA	2 μL
ddH$_2$O	加至 20 μL

37 ℃酶切 6 h。取 5 μL 酶切产物进行 1%琼脂糖凝胶电泳检测。

4.2.2.9 测序

将 PCR、酶切鉴定结果正确的菌液送样测序。每次送 3 个克隆测序,以减少误差。

4.2.2.10 基因及 pCHF3300 载体的酶切回收

酶切体系见 4.2.2.8,基因回收小片段,pCHF3300 回收大片段。回收方法同 2.1.12。回收产物取 3 μL 用 1%琼脂糖凝胶电泳检测,-20 ℃保存。

4.2.2.11 连接体系

将回收的大片段(载体)与小片段(基因)连接,体系见表4-4。

表4-4 连接体系(10 μL)

成分	用量/μL
回收大片段(载体)	1
回收小片段(基因)	5
10×T₄DNA 连接酶 buffer	1
T₄DNA 连接酶	1
ddH₂O	2

16 ℃过夜。连接后转化大肠杆菌 Top10 感受态细胞,提取质粒进行 PCR 和酶切鉴定。转化方法同2.1.13.3,重组质粒的 PCR 鉴定同2.1.13.4,质粒的提取同4.2.2.7,重组质粒的酶切鉴定同4.2.2.8。

4.2.3 *ZmSEC14p* 亚细胞定位载体的构建

植物亚细胞定位载体 pCG3300 - *ZmSEC14p** - eGFP 构建策略如图4-3所示。

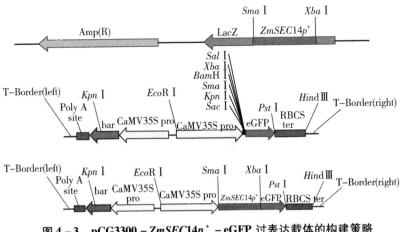

图4-3 pCG3300 - *ZmSEC14p** - eGFP 过表达载体的构建策略

以 $ZmSEC14p^*$ – $pMD^{TM}18$ – T 质粒为模板,设计带有酶切位点的引物,下游引物去掉终止密码子,扩增带有酶切位点 $ZmSEC14p^*$ 的片段。扩增体系与条件同 3.1.19,目的基因的琼脂糖回收同 2.1.12,目的片段和克隆载体的连接同 2.1.13.1,重组质粒的转化与 PCR 鉴定同 2.1.13.3 和 2.1.13.4,小片段(基因)与大片段(载体)的连接同 4.2.2.11,重组质粒的酶切鉴定同 4.2.2.8。

4.2.4 转化拟南芥的研究

4.2.4.1 农杆菌感受态细胞的制备

(1)挑取农杆菌 EHA105 单菌落接种于 5 mL YEB 液体培养基(含 Rif 50 mg/L)中,28 ℃摇菌培养 16 h。

(2)将(1)中培养的农杆菌按 1∶50 比例接种于 50 mL YEB 液体培养基(含 Rif 50 mg/L)中扩摇,于 28 ℃继续培养 6~7 h,分光光度计检测 OD_{600} = 0.5 左右时取出,分装到 1.5 mL 离心管中,冰浴 30 min。

(3)5 000 r/min、4 ℃离心 5 min,在超净工作台中弃去上清液。

(4)每管加入 200 μL 预冷的 0.02 mol/L 氯化钙(含 15% 甘油)重悬菌体, -80 ℃保存。

4.2.4.2 农杆菌转化

(1)取农杆菌感受态细胞置于冰上缓慢融化。

(2)将 2 μL 质粒加入农杆菌感受态细胞中,轻轻混匀。放入液氮冷冻 3 min,37 ℃保温 5 min。

(3)加入 800 μL 无抗生素的 YEB 液体培养基中,28 ℃、200 r/min 振荡培养 4~5 h。

(4)取 200 μL 菌液,涂于 YEB 平板(含 50 mg/L Kana 和 50 mg/L Rif)上,倒置平板于 28 ℃培养约 30 h。待长出明显的菌落后,挑取进行单克隆培养,PCR 用于鉴定阳性菌落,方法同 2.1.13.4。

4.2.4.3 农杆菌介导法转化洋葱表皮细胞

参考刘肖飞等的方法:

(1)将含有 pCG3300 - ZmSEC14p* - eGFP 表达载体的农杆菌接种于 5 mL YEP 液体培养基(含 50 μg/mL Rif 和 50 μg/mL Kana)中,100 r/min、28 ℃ 培养过夜。

(2)按照 1:100 的比例将菌液转移到 50 mL 液体 YEP 培养基(含 50 μg/mL Rif 和 50 μg/mL Kana)中。

(3)取适量菌液,5 000 r/min 离心 15 min,弃去上清液,用含有 100 μmol/L AS 和 100 mmol/L 氯化镁的 MS 液体培养基悬浮菌体。

(4)无菌条件下,用镊子撕取洋葱鳞茎第 5 层内表皮,将其切成若干个 1 cm² 的小块,内侧朝上平铺于 MS 固体培养基上,25 ℃ 暗培养 24 h。

(5)用 MS 液体培养基清洗预培养的洋葱表皮后,将其置于载玻片上,用甘油封片,在激光共聚焦显微镜下观察目的基因的定位情况并拍照保存。

4.2.4.4 蘸花法转化野生型拟南芥

(1)取 10~30 mg 拟南芥种子,先用 70% 乙醇消毒 1 min,再用 2% 次氯酸钠 +0.1% tween -20(表面活性剂,使种子表面完全接触液体)消毒 8 min,可置于振荡器上低速振荡。

(2)用 1 mL 无菌水冲洗 6~8 次。要将次氯酸钠完全洗净,否则将影响萌发率。

(3)将种子浸入无菌水中,放于 4 ℃ 下春化 2~4 d。

(4)在超净工作台上用无菌牙签将种子点播到 MS 固体培养基上,播种后转到人工气候箱中培养,温度为 24 ℃/22 ℃(白天/夜晚),湿度为 60%~70%,光周期为 16 h/8 h(光照/黑暗),光量子通量密度为 150 μmol·m^{-2}·s^{-1}。

(5)待 8~10 d 幼苗长出 4 片真叶后,将幼苗移栽到含有培养基质(草炭土:蛭石:珍珠岩 =10:1:1)的花盆中,相同温度和光照周期继续培养,整个生长期需要添加营养液 3~4 次,每次 1 L。从苗期直至开花,始终保持 1~3 cm 的水层;在开始收籽期,不再需要过多的水分,可保持干燥,此时每隔 3~5 d 加一次水吸足即可。

(6)拟南芥移栽后大约 2 周进入花期(如有需要可剪去主苔,侧苔大约需要 3 周进入花期),待其抽薹至 10~15 cm 时,即刚刚开花,只形成 1~2 个长角果时,可用于转化,此时花蕾状态最佳。转化前将角果及已完全开放的花蕾剪去,仅留下刚刚露白及幼嫩的花蕾。于转化前一天浇足量的水。

(7)挑取单个已鉴定的单菌落于 5 mL YEP 培养基(含 50 mg/L Kan 和 50 mg/L Rif)中,过夜培养后,按 1∶50 的比例加入 500 mL YEB 液体培养基中扩增培养,继续培养 6~8 h,当菌液 $OD_{600}=1.0$ 左右时,8 000 r/min 离心 15 min,收集菌体。

(8)用适量的渗透液(50% MS,5% 蔗糖)充分悬浮菌体至 $OD_{600}=0.8$ 左右,并向含有菌体的渗透液中加入 0.02%(终浓度)的吸附剂 silwet L-77。

(9)将农杆菌菌液倒入培养皿中,将生长状态良好的拟南芥完全浸泡于菌液 1 min,之后用保鲜膜把侵染过的拟南芥包裹好,28 ℃黑暗处理 24 h 后将拟南芥植株转移到正常生长条件下,生长 1 周后再转化 1 次,为了提高转化效率,共转化 2~3 次为宜。

4.2.4.5 转基因植株的筛选

(1)将收获的 T_0 代转基因拟南芥种子置于 37 ℃充分烘干,放于 4 ℃下春化 2~4 d。

(2)春化后的转基因拟南芥种子播种于含有培养基质(草炭土∶蛭石∶珍珠岩=10∶1∶1)的花盆中,待其生长两周左右,用 0.1% 的除草剂喷洒拟南芥,非阳性植株枯萎死亡,阳性植株抗除草剂,可继续生长,对阳性植株进行单株收种。

(3)将收获的 T_1 代拟南芥种子继续播种于培养基质中,取 1~2 片新鲜的拟南芥叶片,提取基因组 DNA,分别进行目的基因与 Bar 基因检测,单株收种。将鉴定阳性的拟南芥种子继续播种于培养基质中,如此直到收获 T_3 代拟南芥种子,选取对除草剂具有 100% 抗性的拟南芥纯合株系进行表型分析与抗逆相关功能的研究。

4.2.4.6 转基因植株基因组 DNA 的提取

(1)取一片拟南芥叶片放入 1.5 mL 离心管中,迅速放入液氮中。

(2)用前端封口的 1 mL 枪头研磨叶片组织,直至研磨成细末。研磨期间常将 1.5 mL 离心管放入液氮中,防止组织融化。

(3)研磨后迅速向每管加入预热的 2×CTAB 提取液 600 μL,放入 60 ℃

水浴 1 h,其间每 10 min 摇动一次,使粉末均匀散布在提取液中。

(4) 加入 600 μL 苯酚 - 三氯甲烷 - 异戊醇 (25∶24∶1),轻轻混匀,室温静置 5 min。

(5) 12 000 ×g 离心 10 min,吸取上清液,加入等体积三氯甲烷 - 异戊醇 (24∶1),轻轻混匀,室温静置 5 min,12 000 ×g 离心 10 min。

(6) 上清液转入另一支新的 1.5 mL 离心管中,加入等体积预冷的异丙醇,-20 ℃ 醇沉 1 h,12 000 ×g 离心 15 min。

(7) 倒掉上清液,沉淀加入 75% 无水乙醇 1 mL 冲洗。12 000 ×g 离心 5 min。此步骤重复两次。

(8) 放入超净工作台,吹干(大约 10 min,不能太干,否则不易溶解)。加入 30 μL ddH$_2$O 溶解,加入 1 μL RNaseA,37 ℃ 保温 1 h,-20 ℃ 保存。

(9) 取 2 μL DNA 用 1% 琼脂糖凝胶电泳进行检测,EB 染色,观察提取质量。

4.2.4.7 转基因植株的 PCR 检测

目的基因的 PCR 反应体系见表 4-5。

表 4-5 目的基因反应体系(25 μL)

成分	用量/μL
DNA	2
10 × PCR buffer	2.5
dNTP(10 mmol/L)	0.5
PCR 引物序列(F)(10 μmol/L)	0.5
PCR 引物序列(R)(10 μmol/L)	0.5
rTaq(5 U/μL)	0.2
ddH$_2$O	18.8

反应程序：

95 ℃ 5 min

95 ℃ 30 s
55 ℃ 30 s } 35 个循环
72 ℃ 60 s

72 ℃ 10 min

4 ℃ 终延伸

4.2.4.8 real–time PCR 分析

RNA 提取同 2.1.7，cDNA 合成同 2.1.9，real–time PCR 反应体系与条件同 2.1.15，引物序列见表 4–6。

表 4–6 real–time PCR 引物序列

基因		引物序列(5'—3')
ZmSEC14p	上游	AAAATCTGCACTTGGTCCTT
	下游	GTCCATCCCGTGAAGTCTAT
RD29A	上游	GGAGGAAATTATTCCACCAGGG
	下游	CAGAATGAGCCGGTGCATCGTG
RD29B	上游	CCTGTCGTGTCTTCTGACCACAC
	下游	CGCTTCCCAGTCCGATGTTTCC
COR6.6	上游	CTGCTGGACAAGGCCAAGGATG
	下游	GGCCGGTCTTGTCCTTCACGAAG
COR15a	上游	AGTGAAACCGCAGATACATTGG
	下游	ACCCTACTTTGTGGCATCCTTAGC
COR47	上游	CCACTACCATCCCGGTACCAGTG
	下游	TTCTCGTCGTGGTGACCAGGAAG
PLC3	上游	CAAGGACATGGGAAGCAACT
	下游	CTTTTGCAAGGGTCGAAGAG
PLC4	上游	AACTTGCTCTGCTCCGTGTT
	下游	AAGAGTGGAACAGCGCGTAT

续表

基因	引物序列(5'—3')	
PLC5	上游	CAAAAGACATGGGAGCCATT
	下游	ACCCGAGAAATCGTCCTTCT
PLC7	上游	CAGGGACTTGGACGATCATT
	下游	TACAGCGTTGGATTTCAGCA
CBF3	上游	TATTTCAGCAAACCATACCAAC
	下游	CTCTAACCTCACAAACCCACTT
CSD1	上游	TTTGAACAGCAGTGAGGGTG
	下游	TCAGTGATTGTGAAGGTGGC
CSD2	上游	GGACCACATTTCAACCCTAACA
	下游	CCATCGGCATTGGCATTT
AtActin1	上游	CCGTGTTGCTCCTGAGGAACATC
	下游	CCTCAGGACAACGGAATCGCTC
AtTIP41	上游	CGCCATACTGTGGAAGTGAA
	下游	CAAAATCGCAAGAGGAGGAA
ZmActin1	上游	CGATTGAGCATGGCATTGTCA
	下游	CCCACTAGCGTACAACGAA
ZmGAPDH	上游	CCCTTCATCACCACGGACTAC
	下游	ACCTTCTTGGCACCACCCT

4.2.4.9 ZmSEC14p 过表达转基因株系的表型鉴定

种子萌发阶段，ZmSEC14p 过表达转基因株系对低温、氯化钠以及 ABA 敏感性的鉴定方法如下：将筛选的纯合转基因拟南芥和野生型株系的种子进行表面消毒，分别点播到含不同浓度的 ABA、氯化钠的 MS 培养基和对照 MS 培养基上，每个株系点 50 粒种子，进行 3 次生物学重复试验。将平板上的种子放到 4 ℃春化 2~4 d 后，再放到正常生长条件[温度为 24 ℃/22 ℃(白天/黑夜)，湿度为 60%~70%，光周期为 16 h/8 h(光照/黑暗)]下萌发，每天统计种子萌发率。将纯合转基因拟南芥和野生型株系的种子点播到正常的 MS

培养基上,并且放置到人工气候箱中进行 4 ℃低温处理,每个株系点 50 粒种子,3 次生物学重复,每隔 5 d 统计种子的萌发率。

(1)幼苗生长阶段:将转基因拟南芥与野生型株系的种子点播于正常的 MS 培养基上,正常生长条件下培养 3 d,待拟南芥完全萌发后,将 3 d 生长一致的幼苗转移到正常的 MS 培养基上并放置到 4 ℃人工气候箱中,试验重复 5 次,处理 15 d 后统计初生根的长度;将 3 d 拟南芥幼苗分别转移到含有 100~200 mmol/L 氯化钠的 MS 培养基上,试验重复 5 次,1 周后统计初生根的长度。

(2)过表达转基因株系的抗冻性鉴定:将用 70% 乙醇和 2% 次氯酸钠消毒后的转基因与野生型种子点播于 MS 培养基上,正常生长 1 周后,将拟南芥幼苗转移到含有培养基质(草炭土∶蛭石∶珍珠岩 = 10∶1∶1)的花盆中。正常生长条件[温度为 24 ℃/22 ℃(白天/黑夜),湿度为 60%~70%,光周期为 16 h/8 h(光照/黑暗)]下生长 3 周后,将温度降低到 -10 ℃,处理 10 h,其他条件不变。处理后在正常生长条件下恢复 1 周,统计存活率。试验进行 3 次生物学重复。

4.2.4.10　O_2^- 和 H_2O_2 组织化学定位

将四周龄的转基因拟南芥与野生型植株分别置于 4 ℃ 和正常生长条件处理 48 h,处理后取其叶片浸泡于 1 mg/mL NBT 溶液(溶于磷酸缓冲液,pH 值为 7.6)中抽真空 20 min,然后室温黑暗培养 3 h,将叶片转移到 80% 的乙醇中,70 ℃ 水浴 10 min 去除叶绿素,然后进行观察。将取下的叶片浸泡于 1 mg/mL 的 DAB 溶液(溶于磷酸缓冲液,pH 值为 7.6)中抽真空 20 min,室温黑暗培养 24 h,将叶片转移到 80% 的乙醇中,70 ℃ 水浴 10 min 去除叶绿素,然后进行观察。

4.2.4.11　*ZmSEC14p* 转基因拟南芥生理指标的测定

SOD 活性的测定方法如下。

(1)植物匀浆液制备:准确称取植物组织质量,按照质量(g)∶体积(mL) = 1∶9 的比例加入匀浆介质(推荐使用 0.1 mol/L 磷酸缓冲液,pH 值为 7.0~7.4),冰水浴下制备成 10% 的组织匀浆液,3 500~4 000 r/min 离心

10 min 后,取上清液进行测定。

(2)按表 4-7 中的步骤进行操作。

表 4-7 SOD 活性测定操作步骤

试剂	测定管	对照管
试剂一/mL	1.0	1.0
样品/mL	a(样品取样量)	—
纯化水/mL	—	b(纯化水取样量)
试剂二/mL	0.1	0.1
试剂三/mL	0.1	0.1
试剂四/mL	0.1	0.1
用旋涡混匀器充分混匀,置 37 ℃ 恒温水浴 40 min		
显色剂/mL	0.1	0.1

混匀,室温放置 10 min,于波长 550 nm 处用 1 cm 光径比色杯,纯化水调零,比色。计算公式:

总 SOD 活性(U/gFW) = (对照管 OD 值 - 总 SOD 测定 OD 值)/对照管 OD 值/0.5 × [反应总体积(mL)/取样量(mL)] × 样品处理前稀释倍数/匀浆液浓度(gFW/mL)

POD 活性测定方法如下。

(1)植物匀浆液的制备同 SOD。

(2)按表 4-8 中步骤进行操作。

表 4-8 POD 活性测定操作步骤

试剂	测定管	对照管
试剂一/mL	2.4	2.4
试剂二应用液/mL	0.3	0.3
试剂三应用液/mL	0.2	—
ddH$_2$O/mL	—	0.2
样本/mL	0.1	0.1
37 ℃ 水浴准确反应 30 min		
试剂四/mL	1.0	1.0

混匀后,3 500 r/min 离心 10 min,取上清液于 420 nm 处用 1 cm 光径比色杯,ddH$_2$O 调零,测定 OD 值。计算公式:

POD 活性(U/mg) = (测定管 OD 值 - 对照管 OD 值)/[12 × 比色光径(1 cm)] × 反应液总体积(mL)/取样量(mL)/反应时间(30 min)/匀浆液浓度(mg/mL) × 1 000

脯氨酸浓度的测定参照张蜀秋的方法。

(1)脯氨酸标准曲线的绘制:取 7 支具塞试管(编号 0~6),按表 4-9 所示加入各试剂,混匀,在沸水中加热 40 min。

表 4-9 各试管中试剂用量

试剂	0	1	2	3	4	5	6
脯氨酸标准溶液/mL	0	0.2	0.4	0.8	1.2	1.6	2.0
H$_2$O/mL	2	1.8	1.6	1.2	0.8	0.4	0
乙酸/mL	2	2	2	2	2	2	2
显色液/mL	3	3	3	3	3	3	3
脯氨酸含量/μg	0	2	4	8	12	16	20

取出试剂冷却,再向各试管中加入 5 mL 甲苯充分振荡,萃取红色物质,静置 40 min 后分层,再用移液器吸取甲苯层,以 0 号管为空白,520 nm 波长下测 OD 值。

以吸光度值为纵坐标、脯氨酸含量为横坐标,绘制标准曲线。

(2)样品测定:取不同处理小麦叶片 0.5 g,剪碎后置于大试管中,加入 3% 磺基水杨酸溶液 5 mL,于沸水浴浸提 10 min。

取出试管后冷却至室温,吸取上清液 2 mL,加 2 mL 乙酸和 3 mL 显色液。于沸水浴加热 40 min,然后再依据标准曲线中的方法进行萃取和比色。

注意:样品颜色程度超过标准曲线中最高浓度颜色时,需要稀释药品。

(3)结果计算:从标准曲线中查出样品中脯氨酸的含量,按下式计算脯氨酸浓度:

脯氨酸浓度[μg/g(鲜或干重)] = $C \times V_0 / V / m \times N$

式中,C 为样品中脯氨酸的含量(μg);V_0 为样品提取液总体积(mL);V 为测定时所吸取的体积（mL）；m 为样品质量（g）；N 为稀释倍数。

PLC 活性的测定原理是根据 PLC 催化水解 NPPC 产生对硝基苯酚，在 410 nm 处有特征吸收峰。

（1）酶液的提取：按照质量（g）：提取液体积（mL）为 1∶10～1∶5 的比例加入提取液（建议称取样品约 0.1 g，加入 1 mL 提取液），冰浴匀浆后于 4 ℃、10 000×g 离心 5 min，取上清液用于测定。提取液缓冲液为 0.25 mol/L Tris-HCl，pH 值为 7.2。

（2）用 Brandford 法进行蛋白质的浓度测定。

（3）反应液的配制：20 mmol/L 的 NPPC 与 60% 山梨醇溶于 0.25 mol/L Tris-HCl 溶液（pH 值为 7.2）中。

（4）5 μL 的酶液提取液与 100 μL 的反应液充分混匀后，37 ℃下反应 30 min，在 410 nm 处测 OD 值。

4.3 结果与分析

4.3.1 *ZmSEC14p* 基因的克隆及表达载体的构建

对基因的 CDS 进行克隆，测序结果表明，克隆的基因包含完整的 ORF。通过对含有目的基因的克隆与表达载体进行酶切、连接、回收、转化等一系列操作，成功构建了 pCHF3300-*ZmSEC14p* 过表达载体（图 4-4）。

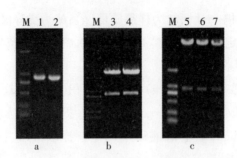

图4-4 pCHF3300-ZmSEC14p 过表达载体的构建结果

1~2. ZmSEC14p 基因的克隆；3~4. ZmSEC14p 克隆载体的酶切鉴定；

5~7. ZmSEC14p 过表达载体的酶切鉴定；M. DL-2000 marker

4.3.2 亚细胞定位载体的构建

对含有目的基因的质粒进行 PCR 鉴定，测序结果表明，克隆的目的基因不含终止密码子，可以与 GFP 基因的 N 末端进行连接。通过对含有目的基因的克隆与表达载体进行酶切、连接、回收、转化等一系列操作，成功构建了 pCG3300-ZmSEC14p*-eGFP 亚细胞定位载体（图4-5）。

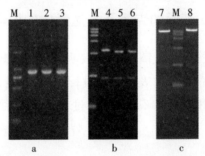

图4-5 pCG3300-ZmSEC14p*-eGFP 亚细胞定位载体的构建结果

1~3. ZmSEC14p* 基因的克隆；4~6. ZmSEC14p* 克隆载体的酶切鉴定；

7~8. ZmSEC14p* 亚细胞定位载体的酶切鉴定；

M. DL-15000 marker、DL-2000 marker

4.3.3 转基因拟南芥纯合株系的筛选及分子鉴定

选取对除草剂具有 100% 抗性的 T_3 拟南芥纯合株系进行目的基因的分子鉴定，结果显示所鉴定的拟南芥植株均含有目的基因，可用于进一步的表型鉴定(图4-6)。

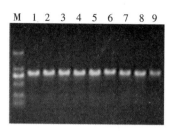

图4-6 部分 T_3 代拟南芥纯合株系目的基因的 PCR 鉴定结果
1~9. 阳性株系；M. DL-2000 marker

4.3.4 非生物胁迫下 *ZmSEC14p* 基因的表达

为了探索 *ZmSEC14p* 基因是否受逆境胁迫诱导表达，试验采用 real-time PCR 的方法检测低温(4 ℃)、盐(250 mmol/L)和 ABA(100 μmol/L)胁迫下玉米叶片和根部 *ZmSEC14p* 基因的表达情况。如图4-7所示，*ZmSEC14p* 冷胁迫下在玉米叶片和根部的表达模式与 ABA 处理相似。冷胁迫2 h 后，*ZmSEC14p* 转录开始上调表达，叶片中4 h 达到高峰，而根部6 h 达到高峰；ABA 处理下，*ZmSEC14p* 的表达在叶片中4 h 达到高峰，根部6 h 达到高峰；盐胁迫下，*ZmSEC14p* 在根部4 h 开始诱导，12 h 后强烈受诱导，在叶片中以较慢速率开始积累，6 h 达到最大，之后下降表达。总之，这些结果表明 *ZmSEC14p* 基因在叶片和根部受冷、盐和 ABA 诱导表达。

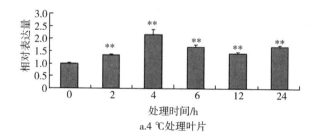

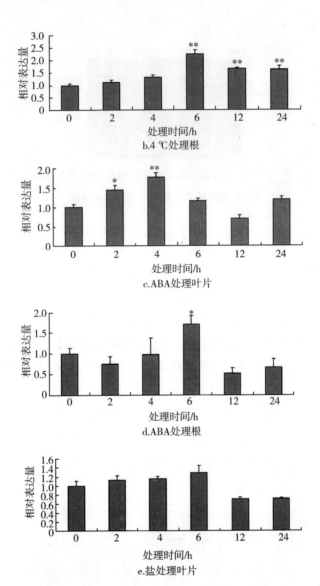

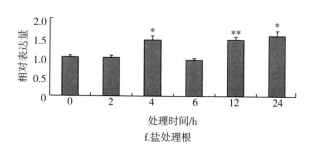

图 4-7 ZmSEC14p 基因的表达分析

注：与 0 h 比较，柱标 * 表示差异显著（$P<0.05$），柱标 ** 表示差异极显著（$P<0.01$），无 * 表示差异不显著（$P>0.05$）。

4.3.5 玉米 ZmSEC14p 组织特异性表达

为了更好地理解 ZmSEC14p 是如何调控发育的生理学过程，利用 real-time PCR 检测不同组织中 ZmSEC14p 基因的表达水平。如图 4-8 所示，ZmSEC14p 在大多数营养和生殖器官的组织中均有表达。ZmSEC14p 在叶片、根、未成熟雄穗以及花粉中表达较高。在所有的组织中，叶片中表达最高，气生根中最低。

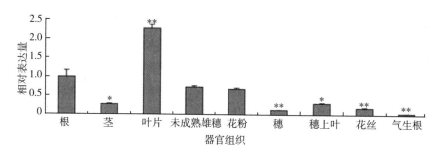

图 4-8 玉米 ZmSEC14p 组织特异性表达

注：与根比较，柱标 * 表示差异显著（$P<0.05$），柱标 ** 表示差异极显著（$P<0.01$），无 * 表示差异不显著（$P>0.05$）。

4.3.6 ZmSEC14p 蛋白的亚细胞定位

尽管 PIPT 在真核细胞的各个组织中普遍存在,不同类型的 PIPT 在发育阶段、组织以及亚细胞定位等方面仍存在不同。为了检测 ZmSEC14p 的亚细胞位置,将 *ZmSEC14p* 基因的 ORF 融合到 GFP 报告基因的 N 端,CaMV35S 启动子控制融合基因的表达。使用农杆菌介导法将重组载体和空载体导入洋葱表皮细胞中,如图 4-9 所示,ZmSEC14p 融合蛋白主要积累于细胞核,而 GFP 在整个细胞中均存在。

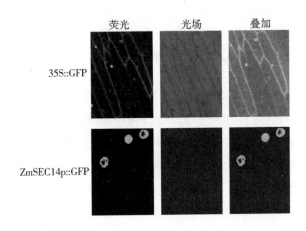

图 4-9 ZmSEC14p∷GFP 亚细胞定位

4.3.7 转基因拟南芥的表型分析

由于 *ZmSEC14p* 基因的表达受 ABA、冷和盐诱导,推测 *ZmSEC14p* 基因可能在这些逆境下影响种子萌发。如图 4-10 所示,在正常温度下,野生型和转基因植株(L2、L3)没有观察到显著的萌发差异。相比于正常的 MS 培养基,野生型和转基因植株的种子在 150 mmol/L 氯化钠培养基上分别推迟了 1 d 和 2 d 萌发。在 150 mmol/L 氯化钠培养基上生长 4 d 后,野生型种子的萌发率为 36.7%,而两个转基因株系的萌发率分别为 20% 和 16.7%。这些结果表明 *ZmSEC14p* 转基因植株在萌发阶段对盐敏感。试验调查了转基因植株的萌发是否受 ABA 影响。野生型和转基因植株在各种浓度的 ABA 培养基上进行萌发测试。5 d 后萌发率如图 4-10b 所示,结果表明 ZmSEC14p 转基因植株在萌发阶段对 ABA 敏感。同时,试验进一步检测了野生型与转基因植株在低温下的种子萌发情况(图 4-10a),4 ℃ 下野生型和转基因植株萌发率差异显著。两个转基因植株在第 5 天开始萌发,第 20 天的萌发率分别是 93% 和 95%,而野生型植株萌发率大约为 75%。这些结果表明 ZmSEC14p 转基因植株在种子萌发阶段对低温不敏感。

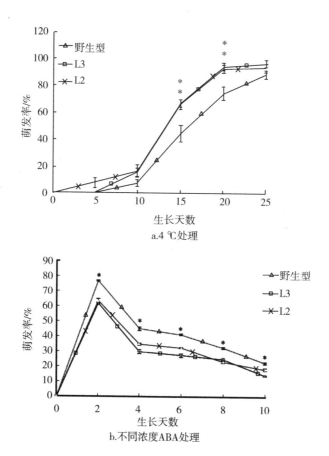

a. 4 ℃ 处理

b. 不同浓度 ABA 处理

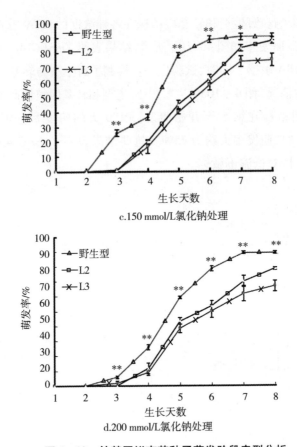

图4-10 转基因拟南芥种子萌发阶段表型分析

注：＊表示差异显著（$P<0.05$），＊＊表示差异极显著（$P<0.01$），无＊表示差异不显著（$P>0.05$）。

盐和冷胁迫下，植株根部生长情况被进一步调查。拟南芥完全萌发后，3 d幼苗被转移到含有150 mmol/L和200 mmol/L氯化钠的MS培养基上，在正常和含不同盐浓度的培养基上，野生型和 *ZmSEC14p* 转基因植株的幼苗生长没有明显差异（图4-11 a和c）。但是，将含有完全萌发拟南芥种子的培养基放到4 ℃，如图4-11 b所示，相比于野生型植株，转基因植株初生根显著伸长。

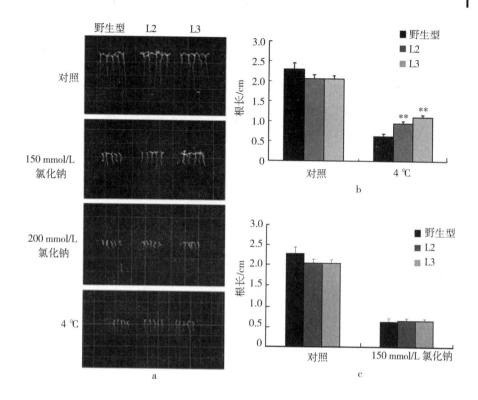

图 4-11 转基因拟南芥植株苗期阶段的表型分析

注：与处理后的野生型相比，柱标 ∗∗ 表示差异极显著（$P<0.01$），无 ∗ 表示差异不显著（$P>0.05$）。

由于 $ZmSEC14p$ 受冷处理诱导，同时在种子萌发与幼苗生长中扮演着重要角色，所以试验进一步调查了 $ZmSEC14p$ 是否能够提高转基因株系的抗冻性。四周龄的野生型与转基因植株 -10 ℃ 处理 10 h，常温下恢复 1 周后统计植株的存活率。冻害胁迫下，转基因植株大约有 67% 的存活率，而野生型植株仅仅有 29% 的存活率（图 4-12）。说明 $ZmSEC14p$ 能够提高转基因植株的抗冻性。

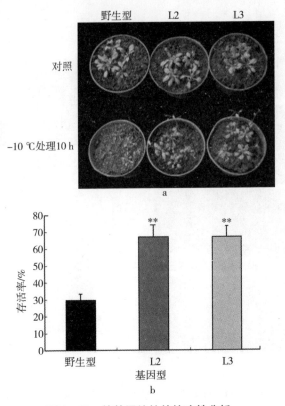

图 4-12 转基因植株的抗冻性分析

注：与野生型相比，柱标 * * 表示差异极显著（$P<0.01$），无 * 表示差异不显著（$P>0.05$）。

4.3.8 ROS 水平检测与抗氧化物分析

有研究表明低浓度的 ROS 能够作为信号分子调节钙离子的信号，钙离子是第二信使，可在低温下调节基因的表达；而高浓度的 ROS 在非生物胁迫下能够导致细胞的氧化损伤。拟南芥过表达 *ZmSEC14p* 能够提高植株的抗冻性。因此，推测 *ZmSEC14p* 可能调控 ROS 的水平。NBT 染色用于 O_2^- 的组织定位，DAB 染色用于 H_2O_2 的组织定位。如图 4-13 所示，常温下，野生型和转基因植株均表现出较低的 ROS 水平，低温处理 2 d 后，相比于野生型植株，O_2^- 和 H_2O_2 的水平在 *ZmSEC14p* 转基因植株叶片中明显降低。

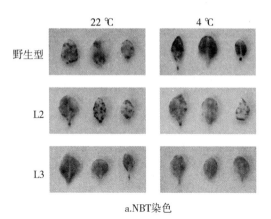

a.NBT染色

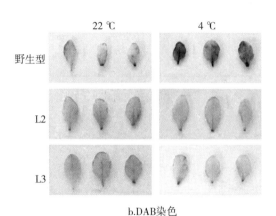

b.DAB染色

图4-13 冷胁迫下ROS水平检测

为了明确冷相关生理指标在提高转基因植株抗冻性过程中是否发生变化,调查了低温处理不同时间下脯氨酸的浓度和抗氧化酶的活性。植物脯氨酸的积累与各种非生物胁迫相关,同时脯氨酸可作为渗透溶剂和ROS的清除剂,保护细胞免遭逆境损害。许多研究报道表明低温驯化过程中,拟南芥脯氨酸的浓度上升。

如图4-14a所示,常温和低温条件下,转基因植株中脯氨酸浓度均显著高于野生型。4℃处理48 h后,两个ZmSEC14p转基因株系的脯氨酸浓度分别是野生型的1.35倍和1.61倍。一些研究表明抗氧化酶在提高植物的抗盐和抗冷过程中起着重要的作用。常温和低温条件下,转基因植株中 SOD 和 POD 的活性均显著高于野生型(图 4-14b 和 c)。研究调查了两个 SOD 基因(*CSD*1 和 *CSD*2)的转录水平,结果表明 *ZmSEC*14*p* 能够在常温和冷处理植株中提高 *CSD*1 和 *CSD*2 的转录水平。

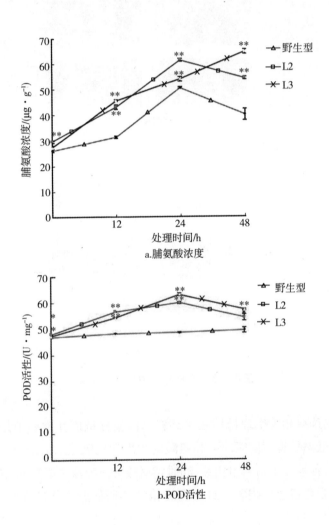

a. 脯氨酸浓度

b. POD 活性

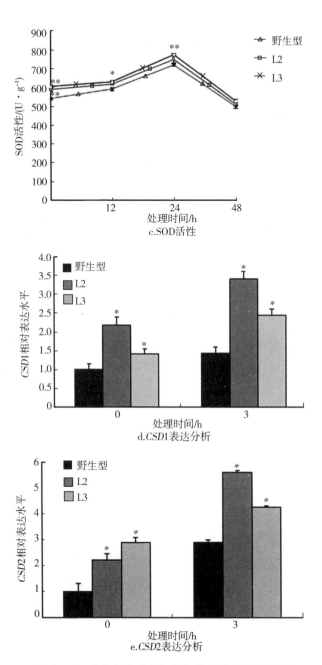

图 4-14 抗氧化物分析与抗氧化关键基因的检测

注：与野生型比较，柱标 * 表示差异显著（$P<0.05$），柱标 * * 表示差异极显著（$P<0.01$），无 * 表示差异不显著（$P>0.05$）。

4.3.9 *PLC* 基因与冷响应关键基因的表达

磷脂可作为第二信号分子合成的前体,在植物响应非生物胁迫中扮演着重要的角色。目前研究最多的切割酶是 PLC,其可通过裂解 PIP2 产生 IP3 和 DAG。IP3 和 DAG 在 PIP 信号转导途径中扮演第二信号分子的角色,其可激活钙离子的释放和蛋白激酶 C。一些研究表明 PI-PLC 活性的抑制能够消除 IP3 的瞬时增加,抑制逆境响应基因 *RD29A* 和 *COR47* 的表达。

过表达 *NbSEC*14 烟草植株能够增加 PLC 和 PLD 的活性,同时激活 JA 信号途径中防御关键基因 *PR-4* 的表达。为了进一步探讨有关 *ZmSEC*14p 介导抗冻性的分子机制,利用 real-time PCR 检测了拟南芥 *PLC* 基因和冷响应关键基因的表达水平。如图 4-15 所示,正常条件下,3 个 *PLC* 基因(*PLC3*、*PLC5*、*PLC7*)在转基因植株中的转录水平显著高于野生型,然而 *PLC4* 和冷响应基因的转录水平在二者之间没有显著差异。冷胁迫下,所有逆境诱导基因在转基因和野生型植株中均上调表达。但是,*PLC3*、*PLC5*、*PLC7*、*CBF3*、*RD29A*、*RD29B*、*COR15a*、*COR6.6* 和 *COR47* 在转基因植株中的表达显著或极显著高于野生型。转基因植株中 *PLC4* 的转录水平高于野生型,但是统计学上没有表现出显著差异。考虑到 *ZmSEC*14p 在转基因植株中调节 *PLC* 基因的表达水平,因此有必要探索冷胁迫下 ZmSEC14p 蛋白和 PLC 活性之间的关系。本研究表明在常温和冷胁迫下 ZmSEC14p 能够提高 PLC 的活性。总之,过表达 *ZmSEC*14p 植株能改变 *PLC* 基因和逆境响应关键基因的表达。

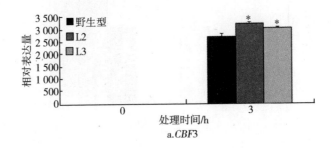

a. *CBF3*

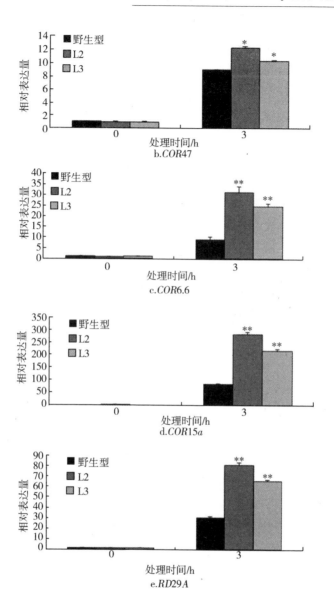

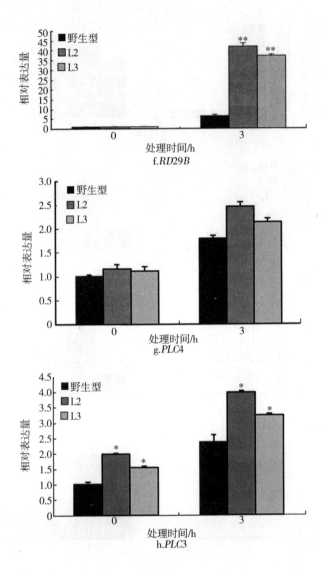

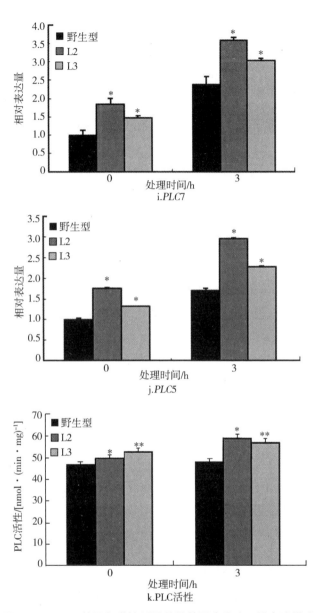

图4-15 PLC基因与逆境诱导关键基因冷胁迫下的表达模式

注:与野生型相比,柱标 * 表示差异显著($P<0.05$),柱标 ** 表示差异极显著($P<0.01$),无 * 表示差异不显著($P>0.05$)。

4.4 讨论

Sec14类蛋白质参与基本生物学过程,例如磷脂代谢、膜转运、极性膜生长、信号转导和逆境响应。本章中,在玉米叶片和根部 *ZmSEC14p* 基因受非生物胁迫诱导表达。组织特异性分析表明 *ZmSEC14p* 基因在叶片中表达最高,气生根中最低。普通小麦孕穗期 *TaSEC14p-5* 基因表达量在茎中最高,根中最低。种子发育过程中,大麦 *HvSec14p* 基因在灌浆期表达量最高,乳熟后期表达量最低;种子萌发过程中,*HvSec14p* 基因在胚根中表达量最高,胚芽鞘里表达量最低。这些结果表明不同Sec14家族成员参与调控不同的发育与生理学过程。

为了验证 *ZmSEC14p* 的生物学功能,试验获得了 *ZmSEC14p* 转基因植株。相比于野生型植株,*ZmSEC14p* 转基因植株在种子萌发阶段对低温不敏感,对ABA和盐胁迫非常敏感。拟南芥 *AtSec14-1* 和 *AtSec14-5* 突变体在萌发阶段对高渗透胁迫和ABA响应非常敏感,原因在于降低了PtdIns磷酸池的水平。因此,未来的工作有必要探讨低温、盐胁迫和ABA处理下PtdIns磷酸池的变化与种子萌发之间的关系。

ROS在植物中扮演双重角色:低浓度下可作为信号分子激活逆境响应,高浓度下会导致不可逆的代谢功能障碍和死亡。非生物胁迫下,ROS清除系统可调控ROS水平。ROS清除系统包括非酶(脯氨酸、GSH)和酶系统(SOD、POD)。脯氨酸可以作为渗透剂保护蛋白质和膜免遭逆境损害以及消除过量的ROS。上升的抗氧化酶活性与提高植物的抗冷、抗盐胁迫能力具有相关性。本章中,过表达 *ZmSEC14p* 能够提高转基因拟南芥植株的抗冻性,提高的抗冻性与脯氨酸浓度的增加、抗氧化酶活性的提高以及降低ROS的水平有关。ICE1-CBF-COR代谢途径在提高植株抗冻性过程中起着至关重要的作用。ICE1只有在低温下诱导 *CBF* 基因的表达。已有研究表明一些胁迫诱导基因能够提高植株的抗逆性。例如,过表达 *CBF/DREB*1、*HVA*1 和 *OSISAP*1 能够提高植株的抗逆性,或者改变逆境响应基因(如 *KIN*1、*COR*15*a* 以及 *RD*29*B*)的表达。A. Kiba等研究表明NbSEC14蛋白可能参与调控茉莉酸依赖性途径防御相关基因的表达。本章中,常温条件下,野生型和转基因植株

具有相似的 *COR* 基因表达水平；冷胁迫后，相比于野生型植株，转基因植株中 *COR* 基因的转录水平显著上升。

生物可基于磷脂的信号级联反应响应各种环境因素和不利生长（生存）条件。例如，在哺乳动物系统中，PLC 产生 IP3 和 DAG，IP3 能够通过结合配体门控的钙离子通道来改变细胞内钙离子的水平，DAG 通过激活一个蛋白激酶 C 来传递信号到下游效应分子。一些植物 Sec14 类蛋白在调节脂信号过程中扮演重要角色。AtSfh1p 可调节 PIP2 的动态平衡，PIP2 是极性膜运输的调节者。AtPATL1，一个新的细胞板相关蛋白，可特异性结合 PIP。有研究人员报道大豆 Sec14p 同源蛋白 Ssh14p 参与渗透信号转导的早期事件和在渗透胁迫下增加 PI-3-kinase 和 PI-4-kinase 的活性。S. Ueno 等使用 RNAi 干扰技术筛选了一个 *ctg-1* 基因，其编码一个 Sec14 类蛋白，敲除此基因增加了对 ROS 的敏感性。SFH2p，酵母 Sec14 同源蛋白家族成员，其功能是通过特异性与巯基过氧化物酶Ⅱ相互作用来调节氧化胁迫。T. Suprunova 等识别了一个干旱诱导基因 *Hsdr*4，抗旱野大麦 Hsdr4 包含一个保守的 SEC14 结构域。T. Krugman 等识别了一个 Sec14 类脂结合蛋白，其在干旱胁迫下抗旱小麦中的表达高于不抗旱小麦。G. Schaaf 等报道了 Sec14 类蛋白刺激了 PIP 的产生。盐、冷以及渗透胁迫可以使植物积累 PtdIns(4,5)P2。PLC 可通过水解 PtdIns(4,5)P2 产生信号分子（如 DAG、PA 和 IP3）。A. Kiba 等发现在青枯菌胁迫下，沉默 *NbSEC*14 能够降低 PLC 的活性，而瞬时过表达 *NbSEC*14 能够激活 PLC 的活性。在盐和渗透胁迫下，PLC 活性的上升可以增加细胞内钙离子水平。许多研究表明非生物胁迫下，PI-PLC 参与和影响各种各样的细胞学过程，例如在大豆中可激活 MAPK 和产生 ROS，高粱中可调节 PEP 羧基酶的基因表达。拟南芥中，PI-PLC、IP3 以及 IP3 门控的钙离子的增加参与调控转录和转录后事件，在响应离子型高渗透胁迫下可以积累脯氨酸渗透调节物质。本章中，相比野生型植株，*ZmSEC14p* 转基因植株可在常温和低温下诱导更高水平 *PLC* 基因的表达，增加 PLC 蛋白的活性。PLC 活性的增加可能促进了脯氨酸含量的提高、抗氧化酶活性的上升以及调节抗冷相关标志基因的表达，进而为 *ZmSEC14p* 能够提高转基因植株抗冷性和抗冻性提供了可能的分子机制。

4.5 结论

(1)*ZmSEC14p* 受低温、ABA 和盐胁迫诱导表达;组织特异性分析表明 *ZmSEC14p* 在玉米叶片中表达最高,气生根中最低。

(2)亚细胞定位试验表明 ZmSEC14p 蛋白主要积累于细胞核。

(3)过表达 *ZmSEC14p* 能够提高转基因植株的抗冻性,主要体现在萌发率的提高、初生根的延长、存活率的提高、脯氨酸浓度的上升、抗氧化酶活性的提高以及由 ROS 造成的氧化损伤的降低方面。

(4)根据上调表达的逆境响应基因推测 *ZmSEC14p* 调节 *PLC* 基因的表达与酶的活性相关。

附录

英文缩写词对照表

英文缩写	英文全称	中文全称及注释
ABA	abscisic acid	脱落酸
ABRE	ABA-responsive element	ABA响应元件
AFP	antifreeze protein	抗冻蛋白
Amp	aminobenzyl penicillin	氨苄青霉素
APS	ammonium persulfate	过硫酸铵
APX	ascorbate peroxidase	抗坏血酸过氧化物酶
AS	acetosyringone	乙酰丁香酮
ATP	adenosine triphophate	三磷酸腺苷
BSA	albumin from bovine serum	牛血清白蛋白
CAT	catalases	过氧化氢酶
CBF	C-repeat binding factors	C-重复结合因子
cDNA-AFLP	cDNA amplified fragments length polymorphism	cDNA扩增片段长度多态性
COG	clusters of orthologous Groups	直系同源簇
COR	cold responsive	冷响应
CRT	C-Repeats	C-重复序列
CSP	cold shock protein	冷休克蛋白
CTAB	cetyl triethylammonium bromide	十六烷基三甲基溴化铵
CTD	C-terminal domain	碳端结构域
DAB	3,3'-Diaminobenzidine	3,3'-二氨基联苯胺
DAG	1,2-Diacylglycerol	1,2-二酰基甘油
DEG	differentially expressed gene	差异表达基因
DEPC	diethylpyrocarbonate	焦碳酸二乙酯
DNA	deoxyribonucleic acid	脱氧核糖核酸
DNase	deoxyribonuclease	脱氧核糖核酸酶
dNTP	deoxynucleotide	脱氧核苷三磷酸

续表

英文缩写	英文全称	中文全称及注释
DREB	dehydration responsive element binding protein	脱水响应元件结合因子
DRE/CRT	dehydration-responsive element/C-repeat	脱水响应元件
EB	ethidium bromide	溴化乙锭
EDTA	ethylenediaminetetraacetic acid	乙二胺四乙酸
ERD	early responsive to dehydration	早期脱水响应
FW	fresh weight	鲜重
GRR1	glucose repression resistance 1	葡萄糖阻遏抗性 1
GSH	glutathione	谷胱甘肽
GST	glutathione S-transferase	谷胱甘肽 S-转移酶
H_2O_2	hydrogen peroxide	过氧化氢
ICE1	inducer of CBF expression 1	CBF 表达诱导因子 1
IP3	inositol-1,4,5-trisphosphate	肌醇-1,4,5-三磷酸
Kana	kanamycin	卡那霉素
KIN	cold induced	冷诱导
LHC II	light harvesting complex II	捕光复合体 II
LRR	leucine-rich repeat protein	亮氨酸富集重复序列蛋白
LTI	low temperature induced	低温诱导
MAPK	mitogen-activated prteinkinase	丝裂原活化蛋白激酶
miRNA	microRNA	微小 RNA
NAC	nascent polypeptide-associated complex	新生多肽相关复合物
NADPH	nicotinamide adenine dinucleotide phosphate	还原型辅酶 II
NahG	SA hydroxylase	SA 羟化酶
NBT	nitroblue tetrazolium	氮蓝四唑
NCBI	national Center for Biotechnology Information	美国国家生物技术信息中心
NMD	nonsense-mediated decay	无义介导衰变
NME	N-terminal met excision	N 末端 Met 切除

续表

英文缩写	英文全称	中文全称及注释
NPC	nuclear pore complex	核孔复合体
NPPC	p-nitrophenylphosphorylcholine	对硝基苯基磷酰胆碱
NUP	nucleoporins	核孔蛋白
O_2^-	superoxide anions	超氧阴离子
ORF	open reading frame	开放阅读框
PAGE	polyacrylamide gel electrophoresis	聚丙烯酰胺凝胶电泳
PA	phosphatidic acid	磷脂酸
PCR	polymerase chain reaction	聚合酶链式反应
PDH45	DNA helicase 45	DNA 解旋酶 45
PEP	phosphoenolpyruvate	磷酸烯醇式丙酮酸
PIP	phosphoinositide	磷酸肌醇
PIPT	phosphatidylinositol transfer protein	磷脂酰肌醇转移蛋白
PIP2	phosphatidylinositol 4,5-bisphophate	磷脂酰肌醇 4,5-二磷酸
PLC	phospholipases C	磷脂酶 C
PLD	phospholipase D	磷脂酶 D
POD	peroxidase	过氧化物酶
PR	pathogenesis-related	发病相关
ProDH	proline dehydrogenase	脯氨酸脱氢酶
PS I	photosystem I	光系统 I
PS II	photosystem II	光系统 II
PtdCho	phosphatidylcholine	磷脂酰胆酸
PtdIns	phosphatidylinositol	磷脂酰肌醇
RAB	responsive to abscisic acid	ABA 响应
RACE	rapid amplification of cDNA end	cDNA 末端快速扩增
RcbA	rubisco activase	Rubisco 活化酶
Rif	rifampicin	利福平
RNA	ribonucleic acid	核糖核酸
ROS	reactive oxygen species	活性氧类

续表

英文缩写	英文全称	中文全称及注释
RT – PCR	reverse transcription polymerase chain reaction	反转录聚合酶链式反应
RuBP	ribulose – 1,5 – bisphosphate	核酮糖 – 1,5 – 二磷酸
SAM	S – adenosyl – L – methionine	脱羧 S – 腺苷 – L – 甲硫氨酸
SAMD	S – adenosyl – L – methionine decarboxylase	S – 腺苷 – L – 甲硫氨酸脱羧酶
SDS	sodium dodecyl sulfate	十二烷基硫酸钠
siRNA	short interfering RNA	短干扰 RNA
SOD	superoxide	超氧化物歧化酶
SR	serine/arginine – rich	丝氨酸/精氨酸富含
TAE	—	核酸电泳缓冲液
TBE	Tris – Borate – EDTA buffer	TBE 缓冲液
TDF	transcript – derived fragment	转录衍生片段
TEMED	N,N,N',N' – tetramethylethylenediamine	N,N,N',N' – 四甲基乙二胺
Tris	trihydroxy aminomethane	三羟甲基氨基甲烷
UQ	ubiqinone	泛醌
UQH2	ubiquinol	还原性泛醌
3' UTR	3' Untranslated region	3' 非翻译区
5' UTR	5' Untranslated region	5' 非翻译区

参考文献

参考文献

[1] SHIFERAW B, PRASANNA B M, HELLIN J, et al. Crops that feed the world 6. Past successes and future challenges to the role played by maize in global food security[J]. Food Security, 2011, 3: 307 – 327.

[2] DELGADO C L. Rising consumption of meat and milk in developing countries has created a new food revolution[J]. The Journal of Nutrition, 2003, 133 (11 Suppl 2): 3907S – 3910S.

[3] 马畅. 吉林省玉米加工业现状研究[J]. 中国农业信息, 2013 (3): 38 – 40.

[4] AGARWAL P K, AGARWAL P, REDDY M K, et al. Role of DREB transcription factors in abiotic and biotic stress tolerance in plants[J]. Plant Cell Reports, 2006, 25(12): 1263 – 1274.

[5] 马树庆, 袭祝香, 王琪. 中国东北地区玉米低温冷害风险评估研究[J]. 自然灾害学报, 2003, 13(3): 137 – 141.

[6] 郝楠. 温度对不同玉米种子萌发及生理特性的影响[D]. 北京: 中国农业科学院, 2011.

[7] 王晓波, 宋凤斌. 玉米非生物逆境生理生态[M]. 北京: 科学出版社, 2005.

[8] 李霞, 李连禄, 王美云, 等. 玉米不同基因型对低温吸胀的响应及幼苗生长分析[J]. 玉米科学, 2008, 16(2): 60 – 65.

[9] DUNCAN W G, HESKETH J D. Net photosynthetic rates, relative leaf growth rates, and leaf numbers of 22 races of 300 maize grown at eight temperatures [J]. Crop Science, 1968, 8(6): 670 – 674.

[10] 史占忠, 贾显明, 张敬涛, 等. 三江平原春玉米低温冷害发生规律及防御措施[J]. 黑龙江农业科学, 2003 (2): 7 – 11.

[11] 余肇福. 作物冷害[M]. 北京: 中国农业出版社, 1991.

[12] 张金龙, 周有佳, 胡敏, 等. 低温胁迫对玉米幼苗抗冷性的影响初探[J]. 东北农业大学学报, 2004, 35(2): 191 – 194.

[13] FAROOQ M, AZIZ T, WAHID A, et al. Chilling tolerance in maize: agronomic and physiological approaches[J]. Crop and Pasture Science, 2009, 60(6): 501 – 516.

[14] 杨德军,洪伟. 低温对玉米生理特性的影响[J]. 民营科技,2011,30(3):120.

[15] 宋立泉. 低温对玉米生长发育的影响[J]. 玉米科学,1997,5(3):58-60.

[16] 王春乙. 东北地区农作物低温冷害研究[M]. 北京:气象出版社,2008.

[17] HUNTER R B, TOLLENAAR M, BREURE C M. Effects of photoperiod and temperature on vegetative and reproductive growth of maize(Zea mays) hybrid[J]. Canadian Journal of Plant Science, 1977, 57(4):1127-1133.

[18] 孙孟梅,姜丽霞,韩俊杰,等. 低温冷害对玉米含水率的影响[J]. 南京气象学院学报,1999,22(4):716-719.

[19] SIDDIQUI K S, CAVICCHIOLI R. Cold-adapted enzymes[J]. Annual Review of Biochemistry, 2006, 75: 403-433.

[20] RUELLAND E, VAULTIER M N, ZACHOWSKI A, et al. Cold signaling and cold acclimation in plants[J]. Advances in Botanical Research, 2009, 49: 35-150.

[21] FURBANK R T, FOYER C H, WALKER D A. Regulation of photosynthesis in isolated spinach chloroplasts during orthophosphate limitation[J]. Biochimica Biophysica Acta, 1987, 894: 552-561.

[22] HURRY V, STRAND A, FURBANK R, et al. The role of inorganic phosphate in the development of freezing tolerance and the acclimatization of photosynthesis to low temperature is revealed by the pho mutants of Arabidopsis thaliana[J]. Plant Journal, 2000, 24(3): 383-396.

[23] GRIFFTH M, ELFMAN B, CAMM E L. Accumulation of plastoquinone a during low temperature growth of winter rye[J]. Plant Physiology, 1984, 74(3): 727-729.

[24] FRACHEBOUD Y, LEIPNER J. The application of chlorophyll fluorescence to study light, temperature, and drought stress[M]//DEELL J R, TOIVONEN P M A. Practical applications of chlorophyll fluorescence in plant biology. Dordrecht: Kluwer Academic Publishers, 2003: 125-150.

[25] KINGSTON-SMITH A H, HARBINSON J, WILLIAMS J, et al. Effect of

chilling on carbon assimilation, enzyme activation, and photosynthetic electron transport in the absence of photoinhibition in maize leaves[J]. Plant Physiology, 1997, 114(3):1039 – 1046.

[26] SONOIKE K. Photoinhibition of photosystem I: Its physiological significance in the chilling sensitivity of plants[J]. Plant and Cell Physiology, 1996, 37(3): 239 – 247.

[27] TJUS S E, MØLLER B L, SCHELLER H V. Photosystem I is an early target of photoinhibition in barley illuminated at chilling temperatures[J]. Plant Physiology, 1998, 116(2): 755 – 764.

[28] TJUS S E, SCHELLER H V, ANDERSSON B, et al. Active oxygen produced during selective excitation of photosystem I is damaging not only to photosystem I, but also to photosystem II [J]. Plant Physiology, 2001, 125(4):2007 – 2015.

[29] ZHANG SUPING, SCHELLER H V. Photoinhibition of photosystem I at chilling temperature and subsequent recovery in *Arabidopsis thaliana* [J]. Plant and Cell Physiology, 2004, 45(11): 1595 – 1602.

[30] HAVAUX M, NIYOGI K K. The violaxanthin cycle protects plants from photooxidative damage by more than one mechanism[J]. The Proceedings of the National Academy of Sciences of the USA, 1999, 96(15): 8762 – 8767.

[31] RUELLAND E, MIGINIAC-MASLOW M. Regulation of chloroplast enzyme activities by thioredoxins: activation or relief from inhibition? [J]. Trends in Plant Science, 1999, 4(4): 136 – 141.

[32] YAN SHUNPING, ZHANG QUNYE, TANG ZHANGCHENG, et al. Comparative proteomic analysis provides new insights into chilling stress responses in rice[J]. Molecular and Cellular Proteomics, 2006, 5(3): 484 – 496.

[33] GOULAS E, SCHUBERT M, KIESELBACH T, et al. The chloroplast lumen and stromal proteomes of *Arabidopsis thaliana* show differential sensitivity to short – and long – term exposure to low temperature[J]. The Plant Journal, 2006, 47(5): 720 – 734.

[34] NAKANO R, ISHIDA H, MAKINO A, et al. In vivo fragmentation of the

large subunit of ribulose - 1,5 - bisphosphate carboxylase by reactive oxygen species in an intact leaf of cucumber under chilling - light conditions[J]. Plant and Cell Physiology, 2006, 47(2): 270 - 276.

[35] DANYLUK J, PERRON A, HOUDE M, et al. Accumulation of an acidic dehydrin in the vicinity of the plasma membrane during cold acclimation of wheat[J]. Plant Cell, 1998, 10(4):623 - 638.

[36] PUHAKAINEN T, HESS M W, MÄKELÄ P, et al. Overexpression of multiple dehydrin genes enhances tolerance to freezing stress in Arabidopsis[J]. Plant Molecular Biology, 2004, 54(5): 743 - 753.

[37] KOAG M C, FENTON R D, WILKENS S, et al. The binding of maize DHN1 to lipid vesicles. Gain of structure and lipid specificity[J]. Plant Physiology, 2003, 131(1): 309 - 316.

[38] BRAVO L A, GALLARDO J, NAVARRETE A, et al. Cryoprotective activity of a cold - induced dehydrin purified from barley[J]. Physiologia Plantarum, 2003, 118(2): 262 - 269.

[39] HARA M, FUJINAGA M, KUBOI T. Radical scavenging activity and oxidative modification of citrus dehydrin[J]. Plant Physiology Biochemistry, 2004, 42(7 - 8): 657 - 662.

[40] HARA M, TERASHIMA S, FUKAYA T, et al. Enhancement of cold tolerance and inhibition of lipid peroxidation by citrus dehydrin in transgenic tobacco[J]. Planta, 2003, 217(2): 290 - 298.

[41] ARTUS N N, UEMURA M, STEPONKUS P L, et al. Constitutive expression of the cold - regulated Arabidopsis thaliana COR15a gene affects both chloroplast and protoplast freezing tolerance[J]. The Proceedings of the National Academy of Sciences of the USA, 1996, 93(23): 13404 - 13409.

[42] ANTIKAINEN M, GRIFFITH M, ZHANG JING, et al. Immunolocalization of antifreeze proteins in winter rye leaves, crowns, and roots by tissue printing[J]. Plant Physiology, 1996, 110(3): 845 - 857.

[43] GRIFFTH M, LUMB C, WISEMAN S B, et al. Antifreeze proteins modify the freezing process in planta[J]. Plant Physiology, 2005, 138(1):

330-340.

[44] GRIFFITH M, YAISH M W F. Antifreeze proteins in overwintering plants: a tale of two activities[J]. Trends in Plant Science, 2004, 9(8): 399-405.

[45] ZHANG DANGQUAN, LIU BING, FENG DONGRU, et al. Significance of conservative asparagine residues in the thermal hysteresis activity of carrot antifreeze protein[J]. Biochemical Journal, 2004, 377: 589-595.

[46] TREMBLAY K, OUELLET F, FOURNIER J, et al. Molecular characterization and origin of novel bipartite cold - regulated ice recrystallization inhibition proteins from cereals[J]. Plant and Cell Physiology, 2005, 46(6): 884-891.

[47] DOXEY A C, YAISH M W F, GRIFFITH M, et al. Ordered surface carbons distinguish antifreeze proteins and their ice - binding regions[J]. Nature Biotechnology, 2006, 24(7): 852-855.

[48] GRIFFTH M, YAISH M W F. Antifreeze proteins in overwintering plants: a tale of two activites[J]. Trends in Plant Science, 2004, 9(8): 399-405.

[49] JIANG WEINING, HOU YAN, INOUYE M. CspA, the major cold - shock protein of *Escherichia coli*, is an RNA chaperone[J]. Journal of Biochemical Chemistry, 1997, 272(1): 196-202.

[50] KIM J Y, PARK S J, JANG B, et al. Functional characterization of a glycine - rich RNA - binding protein 2 in *Arabidopsis thaliana* under abiotic stress conditions[J]. The Plant Journal, 2007, 50(3):439-451.

[51] KIM Y O, KANG H. The role of a zinc finger - containing glycine - rich RNA - binding protein during the cold adaptation process in *Arabidopsis thaliana*[J]. Plant and Cell Physiology, 2006, 47(6):793-798.

[52] UEMURA M, STEPONKUS P L. Effect of cold acclimation on the lipid composition of the inner and outer membrane of the chloroplast envelope isolated from rye leaves[J]. Plant Physiology, 1997, 114(4):1493-1500.

[53] HENDRICKSON L, VLCKOVA A, SELSTAM E, et al. Cold acclimation of the *Arabidopsis dgd*1 mutant results in recovery from photosystem I - limited photosynthesis[J]. FEBS Letters, 2006, 580(20):4959-4968.

[54] IVANOV A G, HENDRICKSON L, KROL M, et al. Digalactosyl – diacylglycerol deficiency impairs the capacity for photosynthetic intersystem electron transport and state transitions in *Arabidopsis thaliana* due to photosytem I acceptor – side limitations [J]. Plant and Cell Physiology, 2006, 47(8): 1146 – 1157.

[55] UEMURA M, JOSEPH R A, STEPONKUS P L. Cold acclimation of *Arabidopsis thaliana* (effect on plasma membrane lipid composition and freeze – induced lesions) [J]. Plant Physiology, 1995, 109(1): 15 – 30.

[56] PALTA J P, WHITAKER B D, WEIS L S. Plasma membrane lipids associated with genetic variability in frezing tolerance and cold acclimation of *Solanum* species [J]. Plant Physiology, 1993, 103(3): 793 – 803.

[57] MIQUEL M, JAMES D J R, DOONER H, et al. *Arabidopsis* requires polyunsaturated lipids for low – temperature survival [J]. The Proceedings of the National Academy of Sciences of the USA, 1993, 90(13): 6208 – 6212.

[58] TASSEVA G, de VIRVILLE J D, CANTREL C, et al. Changes in the endoplasmic reticulum lipid properties in response to low temperature in *Brassica napus* [J]. Plant Physiology Biochemistry, 2004, 42(10): 811 – 822.

[59] GIBSON S, ARONDEL V, IBA K, et al. Cloning of a temperature – regulated gene encoding a chloroplast omega – 3 desaturase from *Arabidopsis thaliana* [J]. Plant Physiology, 1994, 106(4): 1615 – 1621.

[60] ROUGHAN P G. Phosphatidylglycerol and chilling sensitivity in plants [J]. Plant Physiology, 1985, 77(3): 740 – 746.

[61] KAPLAN F, KOPKA J, SUNG D Y, et al. Transcript and metabolite profiling during cold acclimation of *Arabidopsis* reveals an intricate relationship of cold – regulated gene expression with modifications in metabolite content [J]. The Plant Journal, 2007, 50(6): 967 – 981.

[62] NANJO T, KOBAYASHI M, YOSHIBA Y, et al. Antisense suppression of proline degradation improves tolerance to freezing and salinity in *Arabidopsis thaliana* [J]. FEBS Letters, 1999, 461(3): 205 – 210.

[63] KORN M, PETEREK S, MOCK H P, et al. Heterosis in the freezing tolerance, and sugar and flavonoid contents of crosses between *Arabidopsis thaliana* accessions of widely varying freezing tolerance[J]. Plant, Cell and Environment, 2008, 31(6): 813-827.

[64] NIKOLOPOULOS D, MANETAS Y. Compatible solutes and in vitro stability of *Salsola soda* enzymes: proline incompatibility[J]. Phytochemistry, 1991, 30(2): 411-413.

[65] KANDPAL R P, RAO N A. Alterations in the biosynthesis of proteins and nucleic acids in finger millet (*Eleucine coracana*) seedlings during water stress and the effect of proline on protein biosynthesis[J]. Plant Science, 1985, 40(2): 73-79.

[66] VENEKAMP J H. Regulation of cytosol acidity in plants under conditions of drought[J]. Physiologia Plantarum, 2006, 76(1): 112-117.

[67] SCHOBERT B, TSCHESCHE H. Unusual solution properties of proline and its interaction with proteins[J]. Biochimica Biophysica Acta, 1978, 541(2): 270-277.

[68] WYN JONES R G, STOREY R. Betaines[M]//PALEG L G, ASPINAL D. The physiology and biochemistry of drought resistance in plants. New York: Academic Press, 1981: 171-204.

[69] PAPAGEORGIOU G C, MURATA N. The unusually strong stabilizing effects of glycine betaine on the structure and function in the oxygen evolving photosystem II complex[J]. Photosynthesis Research, 1995, 44(3): 243-252.

[70] PRASAD T K, ANDERSON M D, MARTIN B A, et al. Evidence for chilling-induced oxidative stress in maize seedlings and a regulatory role for hydrogen peroxide[J]. Plant Cell, 1994, 6(1): 65-74.

[71] PARK E J, JEKNIC Z, SAKAMOTO A, et al. Genetic engineering of glycine betaine synthesis in tomato protects seeds, plants, and flowers from chilling damage[J]. The Plant Journal, 2004, 40(4): 474-487.

[72] SULPICE R, GIBON Y, CORNIC G, et al. Interaction between exogenous glycine betaine and the photo-respiratory pathway in canola leaf discs[J].

Physiologia Plantarum, 2002, 116(4): 460-467.

[73] IMAI R, ALI A, PRAMANIK H R, et al. A distinctive class of spermidine synthase is involved in chilling response in rice[J]. Journal of Plant physiology, 2004, 161(7):883-886.

[74] COOK D, FOWLER S, FIEHN O, et al. A prominent role for the CBF cold response pathway in configuring the low-temperature metabolome of *Arabidopsis*[J]. The Proceedings of the National Academy of Sciences of the USA, 2004, 101(42): 15243-15248.

[75] PILLAI M A, AKIYAMA T. Differential expression of an S-adenosyl-L-methionine decarboxylase gene involved in polyamine biosynthesis under low temperature stress in japonica and indica rice genotypes[J]. Molecular Genetics and Genomics, 2004, 271(2): 141-149.

[76] KOTZABASIS K, CHRISTAKIS-HAMPSAS M D, ROUBELAKIS-ANGELAKIS K A. A narrow bore HPLC method for the identification and quantitation offree, conjugated, and bound polyamines[J]. Analytical Biochemistry, 1993, 214(2): 484-489.

[77] NAVAKOUDIS E, VRENTZOU K, KOTZABASIS K. A polyamine-and LHC II protease activity-based mechanism regulates the plasticity and adaptation status of the photosynthetic apparatus [J]. Biochimica Biophysica Acta, 2007,1767(4): 261-271.

[78] KIM T E, KIM S K, HAN T J, et al. ABA and polyamines act independently in primary leaves of cold-stressed tomato(*Lycopersicon esculentum*)[J]. Physiologia Plantarum, 2002, 115(3): 370-376.

[79] HE LIXIONG, NADA K, KASUKABE Y, et al. Enhanced susceptibility of photosynthesis to low-temperature photoinhibition due to interruption of chill-induced increase of S-adenosylmethionine decarboxylase activity in leaves of spinach (*Spinacia oleracea* L.) [J]. Plant and Cell Physiology, 2002, 43(2): 196-206.

[80] WI S J, KIM W T, PARK K Y. Overexpression of carnation S-adenosylmethionine decarboxylase gene generates a broad-spectrum tolerance to abiotic

stresses in transgenic tobacco plants[J]. Plant Cell Reports, 2006, 25 (10): 1111-1121.

[81] KASUKABE Y, HE LIXIONG, NADA K, et al. Overexpression of spermidine synthase enhances tolerance to multiple environmental stresses and up-regulates the expression of various stress-regulated genes in transgenic *Arabidopsis thaliana*[J]. Plant and Cell Physiology, 2004, 45(6): 712-722.

[82] GUO ZHENFEI, OU WEN, LU SHULIANG, et al. Differential responses of antioxidative system to chilling and drought in four rice cultivars differing in sensitivity[J]. Plant Physiology and Biochemistry, 2006, 44(11-12): 828-836.

[83] KAPLAN F, GUY C L. Beta-amylase induction and the protective role of maltose during temperature shock[J]. Plant Physiology, 2004, 135(3): 1674-1684.

[84] MÜLLER-MOULÉ P, CONKLIN P, NIYOGI K K. Ascorbate deficiency can limit violaxanthin de-epoxidaseactivity in vivo[J]. Plant Physiology, 2002, 128(3): 970-977.

[85] MIYAKE C, ASADA K. Thylakoid-bound ascorbate peroxidase in spinach chloroplasts and photoreduction of its primary oxidation product monodehydroascorbate radicals in thylakoids[J]. Plant and Cell Physiology, 1992, 33(5): 541-553.

[86] PRICE A, LUCAS P W, LEA P J. Age dependent damage and glutathione metabolism in ozone fumigated barley: a leaf section approach[J]. Journal of Experimental Botany, 1990, 41(231): 1309-1317.

[87] ZHANG ZHIGUO, ZHANG QUAN, WU JINXIA, et al. Gene knockout study reveals that cytosolic ascorbate peroxidase 2(OsAPX2) plays a critical role in growth and reproduction in rice under drought, salt and cold stresses [J]. PLoS One, 2013, 8(2): e57472.

[88] CUI SUXIA, HUANG FANG, WANG JIE, et al. A proteomic analysis of cold stress responses in rice seedlings[J]. Proteomics, 2005, 5(12): 3162-3172.

[89] KWON S J, KWON S I, BAE M S, et al. Role of the methionine sulfoxide reductase MsrB3 in cold acclimation in *Arabidopsis*[J]. Plant and Cell Physiology, 2008, 48(12): 1713-1723.

[90] LLORENTE F, OLIVEROS J C, MARTÍNEZ-ZAPATER J M, et al. A freezing-sensitive mutant of *Arabidopsis*, *frs*1, is a new *aba*3 allele[J]. Planta, 2000, 211(5): 648-655.

[91] MANTYLA E, LANG V, PALVA E T. Role of abscisic acid in drought-induced freezing tolerance, cold acclimation, and accumulation of LT178 and RAB18 proteins in *Arabidopsis thaliana*[J]. Plant Physiology, 1995, 107(1): 141-148.

[92] RABBANI M A, MARUYAMA K, ABE H, et al. Monitoring expression profiles of rice genes under cold, drought, and high-salinity stresses and abscisic acid application using cDNA microarray and RNA gel-blot analyses[J]. Plant Physiology, 2003, 133(4): 1755-1767.

[93] SEKI M, ISHIDA J, NARUSAKA M, et al. Monitoring the expression pattern of around 7,000 *Arabidopsis* genes under ABA treatments using a full-length cDNA microarray[J]. Functional and Integrative Genomics, 2002, 2(6): 282-291.

[94] LANG V, MANTYLA E, WELIN B, et al. Alterations in water status, endogenous abscisic acid content, and expression of *rab*18 gene during the development of freezing tolerance in *Arabidopsis thaliana*[J]. Plant Molecular Biology, 1994, 104(4): 1341-1349.

[95] LEE B H, HENDERSON D A, ZHU JIANKANG. The *Arabidopsis* cold-responsive transcriptome and its regulation by ICE1[J]. Plant Cell, 2005, 17(11): 3155-3175.

[96] de BRUXELLES G L, PEACOCK W J, DENNIS E S, et al. Abscisic acid induces the alcohol dehydrogenase gene in *Arabidopsis*[J]. Plant Physiology, 1996, 111(2): 381-391.

[97] CAPEL J, JARILLO J A, SALINAS J, et al. Two homologous low-temperature-inducible genes from *Arabidopsis* encode highly hydrophobic proteins

[J]. Plant Physiology, 1997, 115(2): 569-576.

[98] KAPLAN F, KOPKA J, HASKELL D W, et al. Exploring the temperature-stress metabolome of *Arabidopsis* [J]. Plant Physiology, 2004, 136(4): 4159-4168.

[99] SCOTT I M, CLARKE S M, WOOD J E, et al. Salicylate accumulation inhibits growth at chilling temperature in *Arabidopsis* [J]. Plant Physiology, 2004, 135(2): 1040-1049.

[100] SASSE J M. Physiological actions of brassinosteroids: an update [J]. Journal of Plant Growth Regulation, 2003, 22(4): 276-288.

[101] HUANG BIN, CHU CHENHUA, CHEN SHULING, et al. A proteomics study of the mung bean epicotyl regulated by brassinosteroids under conditions of chilling stress [J]. Cellular and Molecular Biology Letters, 2006, 11(2): 264-278.

[102] KAGALE S, DIVI U K, KROCHKO J E, et al. Brassinosteroid confers tolerance in *Arabidopsis thaliana* and *Brassica napus* to a range of abiotic stresses [J]. Planta, 2007, 225(2): 353-364.

[103] SHARMA N, CRAM D, HUEBERT T, et al. Exploiting the wild crucifer *Thlaspi arvense* to identify conserved and novel genes expressed during a plant's response to cold stress [J]. Plant Molecular Biology, 2007, 63(2): 171-184.

[104] WONG C E, LI YONG, LABBE A, et al. Transcriptional profiling implicates novel interactions between abiotic stress and hormonal responses in *Thellungiella*, a close relative of *Arabidopsis* [J]. Plant Physiology, 2006, 140(4): 1437-1450.

[105] BELL E, MULLET J E. Characterization of an *Arabidopsis* lipoxygenase gene responsive to methyl jasmonate and wounding [J]. Plant Physiology, 1993, 103(4): 1133-1137.

[106] MISHRS S, KUMAR S, SAHA B, et al. Crosstalk between salt, drought, and cold stress in plants: towards genetic engineering for stress tolerance [M]// TUTEJA N, GILL S S. Abiotic stress responses in plants. Germa-

ny: Wiley – VCH, 2016: 55 –77.

[107] 谢冬微. 冬小麦低温下基因表达谱分析及海藻糖基因家族研究[D]. 哈尔滨:东北农业大学,2014.

[108] LEE I C, HONG S W, WHANG S S, et al. Age – dependent action of an ABA – inducible receptor kinase, RPK1, as a positive regulator of senescence in *Arabidopsis* leaves[J]. Plant and Cell Physiology, 2011, 52(4): 651 –662.

[109] OUYANG SHOUQIANG, LIU YUNFENG, LIU PENG, et al. Receptor – like kinase OsSIK1 improves drought and salt stress tolerance in rice(*Oryza sativa*) plants[J]. The Plant Journal, 2010, 62(2): 316 –329.

[110] de LORENZO L, MERCHAN F, LAPORTE P, et al. A novel plant leucine – rich repeat receptor kinase regulates the response of *Medicago truncatula* roots to salt stress[J]. Plant Cell, 2009, 21(2): 668 –680.

[111] YANG LIANG, JI WEI, ZHU YANMING, et al. GsCBRLK, a calcium/calmodulin binding receptor – like kinase, is a positive regulator of plant tolerance to salt and ABA stress[J]. Journal of Experimental Botany, 2010, 61(9): 2519 –2533.

[112] OSAKABE Y, YAMAGUCHI – SHINOZSKI K, SHINOZAKI K, et al. Sensing the environment: key roles of membrane localized kinases in plant perception and response to abiotic stress[J]. Journal of Experimental Botany, 2013, 64(2): 445 –458.

[113] ABDRAKHAMANOVA A, WANG QIYAN, KHOKHLOVA L, et al. Is microtubule disassembly a trigger for cold acclimation? [J]. Plant and Cell Physiology, 2013, 44(7): 676 –686.

[114] WANG QIYAN, NICK P. Cold acclimation can induce microtubular cold stability in a manner distinct from abscisic acid[J]. Plant and Cell Physiology, 2001, 42(9): 999 –1005.

[115] XIONG LIMING, SCHUMAKER K S, ZHU JIANKANG. Cell signaling during cold, drought, and salt stress[J]. Plant Cell, 2002, 14(Suppl): 165 –183.

[116] ORVAR B L, SANGWAN V, OMANN F, et al. Early steps in cold sensing by plant cells: the role of actin cytoskeleton and membrane fluidity[J]. The Plant Journal, 2000, 23(6): 785-794.

[117] SANGWAN V, FOULDS I, SINGH J, et al. Cold - activation of *Brassica napus* BN115 pomoter is mediated by structural changes in membranes and ctoskeleton, and requires Ca^{2+} influx[J]. The Plant Journal, 2001, 27(1): 1-12.

[118] XIONG LIMING, ISHITANI M, LEE H, et al. The *Arabidopsis LOS5/ABA3* locus encodes a molybdenum cofactor sulfurase and modulates cold stress - and osmotic stress - responsive gene expression[J]. Plant Cell, 2001, 13(9): 2063-2083.

[119] LEE B H, XIONG LIMING, ZHU JIANKANG. A mitochondrial complex Ⅰ defect impairs cold regulated nuclear gene expression[J]. Plant Cell, 2002, 14(6): 1235-1251.

[120] SATOH R, NAKASHIMA K, SEKI M, et al. ACTCAT, a novel cis - acting element for proline - and hypoosmolarity - responsive expression of the *ProDH* gene encoding proline dehydrogenase in *Arabidopsis*[J]. Plant Physiology, 2002, 130(2): 709-719.

[121] OONO Y, SEKI M, NANJO T, et al. Monitoring expression profiles of *Arabidopsis* gene expression during rehydration process after dehydration using ca 7 000 full - length cDNA microarray[J]. The Plant Journal, 2003, 34(6): 868-887.

[122] JEKNI Z, PILLMAN K A, DHILLON T, et al. Hv - CBF2A overexpression in barey accelerates COR gene transcript accumulation and acquisition of freezing tolerance during cold acclimation[J]. Plant Molecular Biology, 2014, 84(1-2): 67-82.

[123] LIU QIANG, KASUGA M, SAKUMA Y, et al. Two transcription factors, DREB1 and DREB2, with an EREBP/AP2 DNA binding domain, separate two cellular signal transduction pathways in drought - and low - temperature responsive gene expression, respectively, in *Arabidopsis* [J]. Plant Cell,

1998, 10(8): 1391 −1406.

[124] NOVILLO F, MEDINA J, SALINAS J. Arabidopsis CBF1 and CBF3 have a different function than CBF2 in cold acclimation and define different gene classes in the CBF regulon[J]. The Proceedings of the National Academy of Sciences of the USA, 2007, 104(52): 21002 −21007.

[125] MARUYAMA K, SAKUMA Y, KASUGA M, et al. Identification of cold − inducible downstream genes of the Arabidopsis DREB1A/CBF3 transcriptional factor using two microarray systems[J]. The Plant Journal, 2004, 38 (16): 982 −993.

[126] CHINNUSAMY V, ZHU JIANHUA, ZHU JIANKANG, et al. Gene regulation during cold acclimation in plants[J]. Physiology Plantarum, 2006, 126(10): 52 −61.

[127] YAMAGUCHI − SHINOZAKI K, SHINOZAKI K. Transcriptional regulatory networks in cellular responses and tolerance to dehydration and cold stresses [J]. Annual Review of Plant Biology, 2006, 57: 781 −803.

[128] STONE J M, PALTA J P, BAMBERG J B, et al. Inheritance of freezing resistance in tuberbearing solanum species: evidence for independent genetic control of nonacclimated freezing tolerance and cold acclimation capacity [J]. The Proceedings of the National Academy of Sciences of the USA, 1993, 90(16): 7869 −7873.

[129] HANNAH M A, WIESE D, FREUND S, et al. Natural genetic variation of freezing tolerance in Arabidopsis[J]. Plant Physiology, 2006, 142(1): 98 −112.

[130] CHINNUSAMY V, OHTA M, KANRAR S, et al. ICE1: a regulator of cold − induced transcriptome and freezing tolerance in Arabidopsis[J]. Genes & Development, 2003, 17(8): 1043 −1054.

[131] NCVILLO F, ALONSO J M, ECKER J R, et al. CBF2/DREB1C is a negative regulator of CBF1/DREB1B and CBF3/DREB1A expression and plays a central role in stress tolerance in Arabidopsis[J]. The Proceedings of the National Academy of Sciences of the USA, 2004, 101(11): 3985 −

3990.

[132] AGARWAL M, HAO YUJIN, KAPOOR A, et al. A R2R3 type MYB transcription factor is involved in the cold regulation of *CBF* genes and in acquired freezing tolerance[J]. The Journal of Biologial Chemistry, 2006, 281(49): 37636 - 37645.

[133] VOGEL J T, ZARKA D G, van BUSKIRK H A, et al. Roles of the CBF2 and ZAT12 transcription factor in configuring the low temperature transcriptome of *Arabidopsis*[J]. The Plant Journal, 2005, 41(2): 195 - 211.

[134] LEE H, GUO YAN, OHTA M, et al. LOS2, a genetic locus required for cold responsive transcription encodes a bi - functional enolase[J]. The EMBO Journal, 2002, 21(11): 2692 - 2702.

[135] DAVIETOVA S, SCHLAUCH K, COUTU J, et al. The zinc - finger protein Zat12 plays a central role in reactive oxygen and abiotic stress signaling in *Arabidopsis*[J]. Plant Physiology, 2005, 139(2): 847 - 856.

[136] MITTLER R, KIM Y S, SONG L, et al. Gain - and loss - of - function mutations in Zat10 enhance the tolerance of plants to abiotic stress[J]. FEBS Letter, 2006, 580(28 - 29): 6537 - 6542.

[137] FOWLER S, THOMASHOW M F. Arabidopsis transcriptome profiling indicates that multiple regulatory pathways are activated during cold acclimation in addition to the CBF cold response pathway[J]. Plant Cell, 2002, 14(8): 1675 - 1690.

[138] KIM J C, LEE S H, CHEONG Y H, et al. A novel cold - inducible zinc finger protein from soybean, SCOF - 1, enhances cold tolerance in transgenic plants[J]. The Plant Journal, 2001, 25(3): 247 - 259.

[139] XIN ZHANGUO, MANDAOKAR A, CHEN JUNPING, et al. *Arabidopsis ESK*1 encodes a novel regulator of freezing tolerance[J]. The Plant Journal, 2007, 49(5): 786 - 799.

[140] ZHU JIANHUA, SHI HUAZHONG, LEE B H, et al. An *Arabidopsis* homeodomain transcription factor gene, *HOS*9, mediates cold tolerance through a CBF - independent pathway[J]. The Proceedings of the National Academy

of Sciences of the USA, 2004, 101(26): 9873-9878.

[141] MASTRANGELO A M, BELLONI S, BARILLI S, et al. Low temperature promotes intron retention in two e-cor genes of durum wheat[J]. Planta, 2005, 221(5): 705-715.

[142] LEE B H, KAPOOR A, ZHU JIANHUA, et al. STABILIZED1, a stress-upregulated nuclearprotein, is required for pre-mRNA splicing, mRNA turnover, and stress tolerance in *Arabidopsis*[J]. Plant Cell, 2006, 18(7): 1736-1749.

[143] PALUSA S G, ALI G S, REDDY A S N. Alternative splicing of pre-mRNAs of *Arabidopsis* serine/arginine-rich proteins: regulation by hormones and stresses[J]. The Plant Journal, 2007, 49(6): 1091-1107.

[144] DONG CHUNHAI, HU XIANGYANG, TANG WEIPING, et al. A putative *Arabidopsis* nucleoporin AtNUP160 is critical for RNA export and required for plant tolerance to cold stress[J]. Molecular Cell Biology, 2006, 26(24): 9533-9543.

[145] COLE C N, SCARCELLI J J. Transport of messenger RNA from the nucleus to the cytoplasm[J]. Current Opinion in Cell Biology, 2006, 18(3): 299-306.

[146] GONG ZHIZHONG, LEE H, XIONG LIMING, et al. RNA helicase-like protein as an early regulator of transcription factors for plant chilling and freezing tolerance[J]. The Proceedings of the National Academy of Sciences of the USA, 2002, 99(17): 11507-11512.

[147] GONG ZHIZHONG, DONG CHUNHAI, LEE H, et al. A DEAD box RNA helicase is essential for mRNA export and important for development and stress responses in *Arabidopsis*[J]. Plant Cell, 2005, 17(1): 256-267.

[148] VASHISHT A A, PRADHAN A, TUTEJA R, et al. Cold- and salinity stress-induced bipolar pea DNA helicase 47 is involved in protein synthesis and stimulated by phosphorylation with protein kinase C[J]. The Plant Journal, 2005, 44(1): 76-87.

[149] SANAN-MISHRA N, PHAM X H, SOPORY S K, et al. Pea DNA heli-

case 45 overexpression in tobacco confers high salinity tolerance without affecting yield[J]. The Proceedings of the National Academy of Sciences of the USA, 2005, 102(2): 509 -514.

[150] CULLEN B R. Nuclear RNA export[J]. Journal of Cell Science, 2003, 116(Pt4): 587 -597.

[151] VERSLUES P E, GUO YAN, DONG CHUNHAI, et al. Mutation of SAD2, an importin beta - domain protein in *Arabidopsis*, alters abscisic acid sensitivity[J]. The Plant Journal, 2006, 47(5): 776 -787.

[152] HIROSE Y, MANLEY J L. RNA polymerase II and the integration of nuclear events[J]. Genes & Development, 2000, 14(12): 1415 -1429.

[153] KOIWA H, BARD A W, XIONG LIMING, et al. C - terminal domain phosphatase - like family members (AtCPLs) differentially regulate *Arabidopsis thaliana* abiotic stress signaling, growth, and development[J]. The Proceedings of the National Academy of Sciences of the USA, 2002, 99 (16): 10893 -10898.

[154] XIONG LIMING, LEE H, ISHITANI M, et al. Repression of stress - responsive genes by FIERY2, a novel transcriptional regulator in *Arabidopsis* [J]. The Proceedings of the National Academy of Sciences of the USA, 2002, 99(16): 10899 -10904.

[155] SUNKAR R, CHINNUSAMY V, ZHU JIANHUA, et al. Small RNAs as big players in plant abiotic stress responses and nutrient deprivation[J]. Trends in Plant Science, 2007, 12(7): 301 -309.

[156] JONES - RHOADES M W, BARTEL D P, BARTEL B. MicroRNAs and their regulatory roles in plants[J]. Annual Review of Plant Biology, 2006, 57: 19 -53.

[157] ZIMMERMANN P, HIRSCH - HOFFMANN M, HENNIG L, et al. *Arabidopsis* microarray database and analysis toolbox [J]. Plant Physiology, 2004, 136(1): 2621 -2632.

[158] SUNKAR R, ZHU JIANKANG. Novel and stress - regulated microRNAs and other small RNAs from *Arabidopsis*[J]. Plant Cell, 2004, 16(8):

2001 - 2019.

[159] SUNKAR R, KAPOOR A, ZHU JIANKANG. Posttranscriptional induction of two Cu/Zn superoxide dismutase genes in *Arabidopsis* is mediated by downregulation of miR398 and important for oxidative stress tolerance[J]. Plant Cell, 2006, 18(8): 2051 - 2065.

[160] LEE H, XIONG LIMING, GONG ZHIZHONG, et al. The *Arabidopsis HOS*1 gene negatively regulates cold signal transduction and encodes a RING finger protein that displays cold - regulated nucleo - cytoplasmic partitioning[J]. Genes & Development, 2001, 15(7): 912 - 924.

[161] DONG CHUNHAI, AGARWAL M, ZHANG YIYUE, et al. The negative regulator of plant cold responses, HOS1, is a RING E3 ligase that mediates the ubiquitination and degradation of ICE1[J]. The Proceedings of the National Academy of Sciences of the USA, 2006, 103(21): 8281 - 8286.

[162] ULRICH H D. Mutual interactions between the SUMO and ubiquitin systems: a plea of no contest[J]. Trends in Cell Biology, 2005, 15(10): 525 - 532.

[163] PRADET - BALADE B, BOULME F, BEUG H, et al. Translation control: bridging the gap between genomics and proteomics?[J]. Trends in Biochemical Science, 2001, 26(4): 225 - 229.

[164] YU LI, YAN JUN, YANG YANJUAN, et al. Overexpression of tomato mitogen - activated protein kinase *SlMPK*3 in tobacco increases tolerance to low temperature stress[J]. Plant Cell, Tissue and Organ Culture, 2015, 121(1): 21 - 34.

[165] THOMASHOW M F. Plant cold acclimation: freezing tolerance genes and regulatory mechanisms[J]. Annual Review of Plant Physiology and Plant Molecular Biology, 1999, 50: 571 - 599.

[166] CHINNUSAMY V, ZHU JIANHUA, ZHU JIANKANG. Cold stress regulation of gene expression in plants[J]. Trends in Plant Science, 2007, 12(10): 444 - 451.

[167] KNIGHT H. Calcium signaling during abiotic stress in plants[J]. Interna-

tional Review of Cytology, 2000, 195: 269-324.

[168] HUANG GOUTAO, MA SHILIANG, BAI LIPING, et al. Signal transduction during cold, salt, and drought stresses in plants[J]. Molecular Biology Reports, 2012, 39(2): 969-987.

[169] STOCKINGER E J, GILMOUR S J, THOMASHOW M F. *Arabidopsis thaliana CBF*1 encodes an AP2 domain - containing transcriptional activator that binds to the C - repeat/DRE, a cis - acting DNA regulatory element that stimulates transcription in response to low temperature and water deficit [J]. The Proceedings of the National Academy of Sciences of the USA, 1997, 94(3): 1035-1040.

[170] KIM Y S, PARK S, GILMOUR S J, et al. Roles of CAMTA transcription factors and salicylic acid in configuring the low - temperature transcriptome and freezing tolerance of *Arabidopsis*[J]. The Plant Journal, 2013, 75(3): 364-376.

[171] PARK S, LEE C M, DOHERTY C J, et al. Regulation of the *Arabidopsis* CBF regulon by a complex low - temperature regulatory network[J]. The Plant Journal, 2015, 82(2): 193-207.

[172] KREPS J A, WU YAJUN, CHANG HURSONG, et al. Transcriptome changes for *Arabidopsis* in response to salt, osmotic, and cold stress[J]. Plant Physiology, 2002, 130(4): 2129-2141.

[173] WANG HAIBO, ZOU ZHURONG, WANG SHASHA, et al. Global analysis of transcriptome responses and gene expression profiles to cold stress of *Jatropha curcas* L. [J]. PLoS One, 2013, 8(12): e82817.

[174] CHENG LIBAO, LI SHUYAN, HUSSAIN J, et al. Isolation and functional characterization of a salt responsive transcription factor, LrZIP from lotus root(*Nelumbo nucifera Gaertn*)[J]. Molecular Biology Reports, 2013, 40(6): 4433-4445.

[175] LIVAK K J, SCHMITTGEN T D. Analysis of relative gene expression data using real - time quantitative PCR and the $2^{-\Delta\Delta C_T}$ method[J]. Methods, 2001, 25(4): 402-408.

[176] 赵继荣,雒淑珍,张增艳,等. cDNA – AFLP 技术及其在植物基因表达分析中的应用[J]. 华北农学报,2009,24(Z1):18 – 22.

[177] 付凤玲,李晚忱. 用 cDNA – AFLP 技术构建基因组转录图谱[J]. 分子植物育种,2003,1(4):523 – 529.

[178] 卢钢,曹家树. cDNA – AFLP 技术在植物表达分析上的应用[J]. 植物学通报,2002,19(1):103 – 108.

[179] LIU BEIBEI, SU SHENGZHONG, WU YING, et al. Histological and transcript analyses of intact somatic embryos in an elite maize (*Zea mays* L.) inbred line Y423[J]. Plant Physiology and Biochemistry, 2015, 92: 81 – 91.

[180] BRUGMANS B, CARMEN A F D, BACHEM C W B, et al. A novel method for the construction of genome wide transcriptome maps[J]. The Plant Journal, 2002, 31(2): 211 – 222.

[181] PERRUC E, CHARPENREAU M, RAMIREZ B C, et al. A novel calmodulin – binding protein functions as a negative regulator of osmotic stress tolerance in *Arabidopsis thaliana* seedlings[J]. The Plant Journal, 2004, 38(3): 410 – 420.

[182] GAO LINLIN, XUE HONGWEI. Global analysis of expression profiles of rice receptor – like kinase genes[J]. Molecular Plant, 2012, 5(1): 143 – 153.

[183] YANG LIANG, WU KANGCHENG, GAO PENG, et al. GsLRPK, a novel cold – activated leucine – rich repeat receptor – like protein kinase from *Glycine soja*, is a positive regulator to cold stress tolerance[J]. Plant Science, 2014, 215 – 216: 19 – 28.

[184] PHILLIPS S E, VINCENT P, RIZZIERI K E, et al. The diverse biological functions of phosphatidylinositol transfer proteins in Eukaryotes[J]. Critical Review in Biochemistry and Molecular Biology, 2006, 41(1): 21 – 49.

[185] CHEN WENQIONG, PROVART N J, GLAZEDROOK J, et al. Expression profile matrix of *Arabidopsis* transcription factor genes suggests their putative functions in response to environmental stresses[J]. Plant Cell, 2002, 14

(3): 559-574.

[186] DAI YIMING, LI WENLI, AN LIJIA. NMD mechanism and the functions of Upf proteins in plant[J]. Plant Cell Reports, 2016, 35(1): 5-15.

[187] UNTERHOLZNER L, IZAURRALDE E. SMG7 acts as a molecular link between mRNA surveillance and mRNA decay[J]. Molecular Cell, 2004, 16(4): 587-596.

[188] GIGLIONE C, MEINNEL T. Organellar peptide deformylases: universality of the N terminal methionine cleavage mechanism[J]. Trends in Plant Science, 2001, 6(12): 566-572.

[189] JEONG H J, SHIN J S, OK S H. Barley DNA-binding methionine aminopeptidase, which changes the localization from the nucleus to the cytoplasm by low temperature, is involved in freezing tolerance[J]. Plant Science, 2011, 180(1): 53-60.

[190] BUKAU B, DEUERLING E, PFUND C, et al. Getting newly synthesized proteins into shape[J]. Cell, 2000, 101(2): 119-122.

[191] BOUCHEREAU A, LARHER F, MARTIN-TANGUY J, et al. Polyamines and environmental challenges: recent development[J]. Plant Science, 1999, 140(2): 103-125.

[192] TABOR C W, TABOR H. Polyamines[J]. Annual Review of Biochemistry, 1984, 53: 749-790.

[193] SHEN WENYUN, NADA K, TACHIBANA S. Involvement of polyamines in the chilling tolerance of cucumber cultivars[J]. Plant Physiology, 2000, 124(1): 431-439.

[194] RHODES D, HANDA S. Amino acid metabolism in relation to osmotic adjustment in plant cells[J]. Environmental Stress in Plants, 1989, 19: 41-62.

[195] HASANUZZAMAN M, HOSSAIN M A, da SILVA J A T, et al. Plant response and tolerance to abiotic oxidative stress: antioxidant defense is a key factor[M]// VENKATESWARLU B, SHANKER A K, SHANKER C, et al. Crop stress and its management: perspectives and strategies. Nether-

lands: Springer, 2012: 261-315.

[196] JOUHET J, MARÉCHAL E, BLOCK M A. Glycerolipid transfer for the building of membranes in plant cells[J]. Progress in Lipid Research, 2007, 46(1): 37-55.

[197] THOLE J M, NIELSEN E. Phosphoinositides in plants: novel functions in membrane trafficking[J]. Current Opinion in Plant Biology, 2008, 11(6): 620-631.

[198] MICHELL R H. Inositol derivatives: evolution and functions[J]. Nature Reviews Molecular Cell Biology, 2008, 9(2): 151-161.

[199] BLOM T, SOMERHARJU P, IKONEN E. Synthesis and biosynthetic trafficking of membrane lipids[J]. Cold Spring Harbor Perspectives in Biology, 2011, 3(8): a004713.

[200] WANG ZHEN, BENNING C. Chloroplast lipid synthesis and lipid trafficking through ER-plastid membrane contact sites[J]. Biochemical Society Transactions, 2012, 40(2): 457-463.

[201] LEV S. Nonvesicular lipid transport from the endoplasmic reticulum[J]. Cold Spring Harbor Perspectives in Biology, 2012, 4(10): a013300.

[202] LEV S. Non-vesicular lipid transport by lipid transfer proteins and beyond [J]. Nature Reviews Molecular Cell Biology, 2010, 11(10): 739-750.

[203] BANKAITIS V A, MALEHORN D E, EMR S D, et al. The *Saccharomyces cerevisiae SEC*14 gene encodes a cytosolic factor that is required for transport of secretory proteins from the yeast Golgi complex[J]. The Journal of Cell Biology, 1989, 108(4): 1271-1281.

[204] BANKAITIS V A, MOUSLEY C J, SCHAAF G. Sec14 superfamily proteins and the crosstalk between lipid signaling and membrane trafficking[J]. Trends in Biochemical Sciences, 2010, 35(3): 150-160.

[205] DAVISON J M, BANKAITIS V A, GHOSH R. Devising powerful genetics, biochemical and structural tools in the functional analysis of phosphatidylinositol transfer proteins (PITPs) across diverse species[J]. Methods Cell Biology, 2012, 108: 249-302.

[206] RITTER A, DITTAMI S M, GOULITQUER S, et al. Transcriptomic and metabolomic analysis of copper stress acclimation in *Ectocarpus siliculosus* highlights signaling and tolerance mechanisms in brown algae[J]. BMC Plant Biology, 2014, 14: 116.

[207] LEE J, LIM Y P, HAN C T, et al. Genome-wide expression profiles of contrasting inbred lines of Chinese cabbage, Chiifu and Kenshin under temperature stress[J]. Genes & Genomics, 2013, 35(3): 273-288.

[208] CLEVES A E, MCGEE T, BANKAITIS V. Phospholipid transfer proteins: a biological debut[J]. Trends in Cell Biology, 1991, 1(1): 31-34.

[209] ARAVIND L, NEUWALD A F, PONTING C P. Sec14p-like domains in NF1 and Dbl-like proteins indicate lipid regulation of Ras and Rho signaling[J]. Current Biology, 1999, 9(6): R195-R197.

[210] KIBA A, NAKANO M, VINCENT-POPE P, et al. A novel Sec14 phospholipid transfer protein from *Nicotiana benthamiana* is up-regulated in response to *Ralstonia solanacearum* infection, pathogen associated molecular patterns and effector molecules and involved in plant immunity[J]. Journal of Plant Physiology, 2012, 169(10): 1017-1022.

[211] NILE A H, BANKAITIS A V, GRABON A. Mammalian diseases of phosphatidylinositol transfer proteins and their homologs[J]. Journal of Clinical Lipidology, 2010, 5(6): 867-897.

[212] ALLEN-BAUME V, SÉGUI B, COCKCROFT S. Current thoughts on the phosphatidylinositol transfer protein family[J]. FEBS Letter, 2002, 531(1): 74-80.

[213] COCKCROFT S, GARNER K. Function of the phosphatidylinositol transfer protein gene family: is phosphatidylinositol transfer the mechanism of action? [J]. Critical Reviews in Biochemistry and Molecular Biology, 2011, 46(2): 89-117.

[214] GARNER K, HUNT A N, KOSTER G, et al. Phosphatidylinositol transfer protein, cytoplasmic 1 (PITPNC1) binds and transfers phosphatidic acid [J]. Journal of Biological Chemistry, 2012, 287(38): 32263-32276.

[215] MOUSLEY C J, TYERYAR K R, VINCENT-POPE P, et al. The Sec-14-superfamily and the regulatory interface between phospholipid metabolism and membrane trafficking[J]. Biochimica et Biophysica Acta, 2007, 1771(6): 727-736.

[216] DOVE S K, COOKE F T, DOUGLAS M R, et al. Osmotic stress activates phosphatidylinositol-3,5-bisphosphate synthesis[J]. Nature, 1997, 390(6656): 187-192.

[217] MEIJER H J G, DIVECHA N, van den ENDE H, et al. Hyperosmotic stress induces rapid synthesis of phosphatidyl-D-inositol 3,5-bisphosphate in plant cells[J]. Planta, 1999, 208(2): 294-298.

[218] KIELBOWICZ-MATUK A, BANACHOWICZ E, TURSKA-TARSKA A, et al. Expression and characterization of a barley phosphatidylinositol transfer protein structurally homologous to the yeast Sec14p protein[J]. Plant Science, 2016, 246: 98-111.

[219] KEARNS M A, MONKS D E, FANG M, et al. Novel developmentally regulated phosphoinositide binding proteins from soybean whose expression bypasses the requirement for an essential phosphatidylinositol transfer protein in yeast[J]. EMBO Journal, 1998, 17(14): 4004-4017.

[220] 苏世超,唐益苗,徐磊,等. 普通小麦 TaSEC14p-5 基因的克隆及表达分析[J]. 农业生物技术学报,2016,24(8):1129-1137.

[221] 刘肖飞,梁卫红. 根癌农杆菌介导的 GFP 在洋葱表皮细胞定位研究[J]. 河南师范大学学报(自然科学版),2009,37(1):123-125,150.

[222] 张蜀秋. 植物生理学实验技术教程[M]. 北京:科学出版社,2011.

[223] KURIOKA S, MATSUDA M. Phospholipase C assay using p-nitrophenylphosphorylcholine together with sorbitol and its application to studying the metal and detergent requirent of the enzyme[J]. Analytical Biochemistry, 1976, 75(1): 281-289.

[224] PRICE A H, TAYLOR A, RIPLEY S J. Oxidative signals in tobacco increase cytosolic calcium[J]. Plant Cell, 1994, 6(9): 1301-1310.

[225] CATALÁ R, SANTOS E, ALONSO J M, et al. Mutations in the Ca^{2+}/H^+

transporter CAX1 increase *CBF/DREB*1 expression and the cold acclimation response in *Arabidopsis*[J]. Plant Cell, 2003, 15(12): 2940-2951.

[226] KNIGHT H, TREWAVAS A J, KNIGHT M R. Cold calcium signaling in *Arabidopsis* involves two cellular pools and a charge in calcium signature after acclimation[J]. Plant Cell, 1996, 8(3): 489-503.

[227] MONROY A F, DHINDSA R S. Low-temperature signal transduction: induction of cold acclimation-specific genes of alfalfa by calcium at 25 degrees C[J]. Plant Cell, 1995, 7(3): 321-331.

[228] REDDY A R, CHAITANYA K V, VIVEKANANDAN M. Drought-induced responses of photosynthesis and antioxidant metabolism in higher plants[J]. Journal of Plant Physiology, 2004, 161(11): 1189-1202.

[229] XU WEIRONG, JIAO YUNTONG, LI RUIMIN, et al. Chinese wild-growing *Vitis amurensis ICE*1 and *ICE*2 encode *MYC*-type *bHLH* transcription activators that regulate cold tolerance in *Arabidopsis*[J]. PLoS One, 2014, 9(7): e102303.

[230] SHAN DAPENG, HUANG JINGUANG, YANG YUTAO, et al. Cotton GhDREB1 increases plant tolerance to low temperature and is negatively regulated by gibberellic acid[J]. New Phytologist, 2007, 176(1): 70-81.

[231] HUANG JINGUANG, YANG MEI, LIU PEI, et al. *GhDREB*1 enhances abiotic stress tolerance, delays GA-mediated development and represses cytokinin signalling in transgenic *Arabidopsis*[J]. Plant Cell and Environment, 2009, 32(8): 1132-1145.

[232] ZONG XIAOJUAN, LI DAPENG, GU LINGKUN, et al. Abscisic acid and hydrogen peroxide induce a novel maize group CMAP kinase gene, *ZmMPK7*, which is responsible for the removal of reactive oxygen species [J]. Planta, 2009, 229(3): 485-495.

[233] NING JING, LI XIANGHUA, HICKS L M, et al. A Raf-like MAPKKK gene *DSM*1 mediates drought resistance through reactive oxygen species scavenging in rice[J]. Plant Physiology, 2010, 152(2): 876-890.

[234] VOSSEN J H, ABD - EI - HALIEM A, FRADIN E F, et al. Identification of tomato phosphatidylinositol - specific phospholipase - C (PI - PLC) family members and the role of PLC4 and PLC6 in HR and disease resistance [J]. The Plant Journal, 2010, 62(2): 224 - 239.

[235] TASMA I M, BRENDEL V, WHITHAM S A, et al. Expression and evolution of the phosphoinositide - specific phospholipase C gene family in *Arabidopsis thaliana* [J]. Plant Physiology and Biochemistry, 2008, 46(7): 627 - 637.

[236] de WALD D B, TORABINEJAD J, JONES C A, et al. Rapid accumulation of phosphatidylinositol 4, 5 - bisphosphate and inositol 1, 4, 5 - trisphosphate correlates with calcium mobilization in salt - stressed *Arabidopsis* [J]. Plant Physiology, 2001, 126(2): 759 - 769.

[237] TAKAHASHI S, KATAGIRI T, HIRAYAMA T, et al. Hyperosmotic stress induced a rapid and transient increase in inositol 1, 4, 5 - trisphosphate independent of abscisic acid in *Arabidopsis* cell culture [J]. Plant and Cell Physiology, 2001, 42(2): 214 - 222.

[238] KEOGH M R. New insights into phospholipid metabolism and signaling in plants [D]. Raleigh: NC State University, 2009.

[239] HARE P D, CRESS W A. Metabolic implications of stress - induced proline accumulation in plants [J]. Plant Growth Regulation, 1997, 21(2): 79 - 102.

[240] TROVATO M, MATTIOLI R, COSTANTINO P. Multiple roles of proline in plant stress tolerance and development [J]. Rendiconti Lincei, 2008, 19(4): 325 - 346.

[241] GILL S S, TUTEJA N. Reactive oxygen species and antioxidant machinery in abiotic stress tolerance in crop plants [J]. Plant Physiology and Biochemistry, 2010, 48(12): 909 - 930.

[242] BROWSE J, XIN ZHANGUO. Temperature sensing and cold acclimation [J]. Current Opinion in Plant Biology, 2001, 4(3): 241 - 246.

[243] DUBOUZET J G, SAKUMA Y, ITO Y, et al. *OsDREB* genes in rice, *Ory-*

za sativa L. , encode transcription activators that function in drought –, high – salt – and cold – responsive gene expression[J]. The Plant Journal, 2003, 33(4): 751 –763.

[244] MUKHOPADHYAY A, VIJ S, TYAGI A K. Overexpression of a zinc finger protein gene from rice confers tolerance to cold, dehydration, and salt stress in transgenic tobacco[J]. The Proceedings of the National Academy of Sciences of the USA, 2004, 101(16): 6309 –6314.

[245] ABE H, URAO T, ITO T, et al. *Arabidopsis* AtMYC2 (bHLH) and AtMYB2(MYB)function as transcriptional activators in abscisic acid signaling [J]. Plant Cell, 2003, 15(1): 63 –78.

[246] GILMOUR S J, FOWLER S G, THOMASHOW M F. *Arabidopsis* transcriptional activators CBF1, CBF2, and CBF3 have matching functional activities[J]. Plant Molecular Biology, 2004, 54(5): 767 –781.

[247] SAKAMOTO H, MARUYAMA K, SAKUMA Y, et al. *Arabidopsis* Cys2/His2 – type zinc – finger proteins function as transcription repressors under drought, cold, and high – salinity stress conditions[J]. Plant Physiology, 2004, 136(1): 2734 –2746.

[248] SINGH A, BHATNAGAR N, PANDEY A, et al. Plant phospholipase C family: regulation and functional role in lipid signaling[J]. Cell Calcium, 2015, 58(2): 139 –146.

[249] VINCENT P, CHUA M, NOGUE F, et al. A Sec14p nodulin domain phosphatidylinositol transfer protein polarizes membrane growth of *Arabidopsis thaliana* root hairs[J]. The Journal of Cell Biology, 2005, 168(5): 801 –812.

[250] PETERMAN T K, OHOL Y M, MCREYNOLDS L J, et al. Patellin1, a novel Sec14 – like protein, localizes to the cell plate and binds phosphoinositides[J]. Plant Physiology, 2004, 136(2): 3080 –3094.

[251] MONKS D E, AGHORAM K, COURTNEY P D, et al. Hyperosmotic stress induces the rapid phosphorylation of a soybean phosphatidylinositol transfer protein homolog through activation of the protein kinases SPK1 and SPK2

[J]. Plant Cell, 2001, 13(5): 1205 – 1219.

[252] UENO S, YASUTAKE K, TOHYAMA D, et al. Systematic screen for genes involved in the regulation of oxidative stress in the nematode *Caenorhabditis elegans*[J]. Biochemical and Biophysical Reserach Communications, 2012, 420(3): 552 – 557.

[253] CHA M K, HONG S K, OH Y M, et al. The protein interaction of *Saccharomyces cerevisiae* cytoplasmic thiol peroxidase II with Sfh2p and its in vivo function [J]. Journal of Biological Chemistry, 2003, 278 (37): 34952 – 34958.

[254] SUPRUNOVA T, KRUGMAN T, DISTELFELD A, et al. Identification of a novel gene (*Hsdr4*) involved in water stress tolerance in wild barley [J]. Plant Molecular Biology, 2007, 64(1 – 2): 17 – 34.

[255] KRUGMAN T, CHAGUÉ V, PELEG Z, et al. Multilevel regulation and signaling processes associated with adaptation to terminal drought in wild emmer wheat [J]. Functional & Integrative Genomics, 2010, 10 (2): 167 – 186.

[256] SCHAAF G, ORTLUND E A, TYERYAR K R, et al. Functional anatomy of phospholipid binding and regulation of phosphoinositide homeostasis by proteins of the Sec14 superfamily [J]. Molecular Cell, 2008, 29 (2): 191 – 206.

[257] SMOLENSKA – SYM G, KACPERSKA A. Phosphatidylinositol metabolism in low temperature – affected winter oilseed rape leaves [J]. Physiologia Plantarum, 1994, 91(1): 1 – 8.

[258] PICAL C, WESTERGREN T, DOVE S K, et al. Salinity and hyperosmotic stress induce rapid increases in phosphatidylinositol 4, 5 – bisphosphate, diacylglycerol pyropho sphate, and phosphatidylcholine in *Arabidopsis tha-liana* cells [J]. Journal of Biological Chemistry, 1999, 274 (53): 38232 – 38240.

[259] IM J H, LEE H, KIM J, et al. Soybean MAPK, GMK1 is dually regulated by phosphatidic acid and hydrogen peroxide and translocated to nucleus du-

ring salt stress[J]. Molecules and Cells, 2012, 34(3): 271-278.
[260] MONREAL J A, ARIAS - BALDRICH C, PEREZ - MONTANO F, et al. Factors involved in the rise of phosphoenolpyruvate carboxylase - kinase activity caused by salinity in sorghum leaves[J]. Planta, 2013, 237(5): 1401-1413.
[261] PARRE E, GHARS M A, LEPRINCE A S, et al. Calcium signaling via phospholipase C is essential for proline accumulation upon ionic but not nonionic hyperosmotic stresses in *Arabidopsis*[J]. Plant Physiology, 2007, 144(1): 503-512.